Springer Series in **Nonlinear Dynamics**

ND Springer Series in **Nonlinear Dynamics**

Series Editors: F. Calogero, B. Fuchssteiner, G. Rowlands, M. Wadati, and V. E. Zakharov

Solitons – Introduction and Applications
Editor: M. Lakshmanan

What Is Integrability?
Editor: V.E. Zakharov

Rossby Vortices and Spiral Structures
By M.V. Nezlin and E. N. Snezhkin

Algebro-Geometrical Approach to Nonlinear Evolution Equations
By E. D. Belokolos, A. I. Bobenko, V. Z. Enolsky, A.R. Its and V. B. Matveev

Darboux Transformations and Solitons
By V. B. Matveev and M. A. Salle

Optical Solitons
By F. Abdullaev, S. Darmanyan and P. Khabibullaev

Wave Turbulence Under Parametric Excitation
Applications to Magnetics
By V. S. L'vov

Kolmogorov Spectra of Turbulence I Wave Turbulence
By V. E. Zakharov, V. S. L'vov and G. Falkovich

Nonlinear Processes in Physics
Editors: A. S. Fokas, D. J. Kaup, A. C. Newell and V. E. Zakharov

Important Developments in Soliton Theory
Editors: A. S. Fokas and V. E. Zakharov

F. Abdullaev S. Darmanyan
P. Khabibullaev

Optical Solitons

With 34 Figures

Springer-Verlag
Berlin Heidelberg GmbH

Professor Dr. Fatkhulla Abdullaev
Dr. Sergei Darmanyan

Physical-Technical Institute, Theoretical Division, Timiryazeva (G. Mavlyanova) Str. 2b,
700084 Tashkent – 84, Uzbekistan, CIS

Professor Dr. Pulat Khabibullaev
Thermal Physics Department, 700135 Tashkent – 135, Uzbekistan, CIS

Volume Editor

Professor Vladimir E. Zakharov

Landau Institute for Theoretical Physics, Russian Academy of Sciences, ul. Kosygina 2,
117334 Moscow, Russia, CIS

Guest Editor

Professor Juri Engelbrecht

Institute of Cybernetics, Estonian Academy of Sciences, Akadeemia tee 21
EE-0026 Tallinn, Estonia

ISBN 978-3-642-87718-6   ISBN 978-3-642-87716-2 (eBook)
DOI 10.1007/978-3-642-87716-2

Library of Congress Cataloging-in-Publication Data. Abdullaev, F. Kh. (Fatkhulla Khabibullaevich) [Opticheskie
solitony. English] Optical solitons / F. Kh. Abdullaev, S. A. Darmanyan, P. K. Khabibullaev. p. cm. – (Springer series
in nonlinear dynamics) Translation of: Opticheskie solitony. Includes bibliographical references and index.
ISBN 978-3-642-87718-6 (New York: alk. paper) 1. Solitons. 2. Nonlinear optics.
I. Darmanian, S. A. (Sergeĭ Arakelovich) II. Bar'iakhtar, Viktor Grigor'evich. III. Title. IV. Series.
QC174.26.W28A2312   1993   535'.2–dc20   93-13774

Typesetting: Springer T$_E$X in-house system
57/3140 - 5 4 3 2 1 0 - Printed on acid-free paper

Preface

The investigation of nonlinear wave phenomena has been one of the main directions of research in optics for the last few decades. Soliton concepts applied to the description of intense electromagnetic beams and ultrashort pulse propagation in various media have contributed much to this field. The notion of solitons has proved to be very useful in describing wave processes in hydrodynamics, plasma physics and condensed matter physics. Moreover, it is also of great importance in optics for ultrafast information transmission and storage, radiation propagation in resonant media, etc.

In 1973, Hasegawa and Tappert made a significant contribution to optical soliton physics when they predicted the existence of an envelope soliton in the regime of short pulses in optical fibres. In 1980, Mollenauer et al. conducted experiments to elucidate this phenomenon. Since then the theory of optical solitons as well as their experimental investigation has progressed rapidly. The effects of inhomogeneities of the medium and energy pumping on optical solitons, the interaction between optical solitons and their generation in fibres, etc. have all been investigated and reported. Logical devices using optical solitons have been developed; new types of optical solitons in media with Kerr-type nonlinearity and in resonant media have been described.

This book presents the theory of optical solitons and its application. We also consider a number of related problems in resonant media and the problem of quantum optical solitons. The book has been written both as an introduction to the theory (especially Chap. 2), and as a monograph containing the latest advances in this area. This book will be helpful to scientists and students with an interest in this fascinating and rapidly developing field of nonlinear optics.

Tashkent, July 1993
F. Abdullaev
S. Darmanyan
P. Khabibullaev

Contents

1. **Introduction** 1

2. **Short Pulses in Optical Fibres** 5
   2.1   Solitons in Single-Mode Fibres 6
   2.2   Self-Phase Modulation of Pulses 11
   2.3   Soliton Regime of Ultrashort Pulse Propagation.
         Bright and Dark Solitons 13
         2.3.1   Presoliton Region 16
         2.3.2   Near-Soliton Region 17
         2.3.3   Multi-Soliton Region 18
   2.4   Observations of Solitons 21
   2.5   Electromagnetic Shock Waves 23
   2.6   Influence of Higher Order Dispersion and Dissipation
         on Soliton Dynamics 27
   2.7   Amplification of Optical Solitons in Fibres 30
   2.8   Modulation Instability of Electromagnetic Waves 37
   2.9   Generation of Periodic Pulse Sequences 39
   2.10 Multi-Mode Effects on Soliton Regimes of Propagation ... 46
   2.11 The Soliton Laser 48
   2.12 The Soliton Self-Frequency Shift 52
   2.13 Multi-Component (Vector) Optical Solitons 54

3. **Soliton Interaction** 57
   3.1   General Representations
         in Integrable and Nonintegrable Systems:
         Direct and IST-Based Methods 57
   3.2   Interaction Between Optical Solitons 62
   3.3   Inelastic Soliton Interactions 66
   3.4   Interaction Between a Soliton and a Nonlinear Periodic Wave
         in an Optical Fibre 71
   3.5   Solitons in Two Tunnel-Coupled Optical Waveguides 76

4. **Statistical Dynamics of Optical Solitons** 81
   4.1   Random Pulses in an Optical Fibre.
         Numerical Simulation in the Presoliton Region 81

4.2   Noise Signals in the Near Field   86
4.3   Noise Signals in the Far Field   88
4.4   Random Pulses in Fibres (Soliton Region)   91
4.5   Solitons in Random Nonuniform Fibres   97
4.6   Random Amplification of Solitons   101
4.7   Dynamic Chaos of Optical Solitons   105
4.8   Optical Turbulence in Passive Optical Resonators   110

5.   Optical Solitons in Resonant and Active Media   115
5.1   The Maxwell-Bloch System. Soliton Solutions   115
5.2   Optical Solitons in an Active Fibre   119
5.3   Scattering of a Weak Wave on an Optical Soliton   121
5.4   Theory of Superfluorescence   125
5.5   Solitons in Stimulated Raman Scattering   130
5.6   The Evolution of SRS Solitons
      Under the Action of Molecular Relaxation   133

6.   Quantum Optical Solitons   139
6.1   Hamiltonian Formulation of the IST-Method
      and Classical r-Matrix   140
6.2   Quantum Nonlinear Schrödinger Equation (QNLSE)   144
6.3   Quantum Sine-Gordon System   149
6.4   The Dicke Model   152

7.   Conclusion   159

Appendix   161

References   181

Subject Index   189

1. Introduction

Soliton concepts have deeply penetrated into different branches of physics. This process started in hydrodynamics at the end of the last century, was extended to plasma physics in the fifties and then was later applied to solid state physics. Further investigations indicated the universal nature of excitations of this kind in nonlinear media.

Nonlinear optics has not been excluded from these developments. In the early sixties the role of nonlinear stationary wave packets in various processes involving intense electromagnetic fields was noted. Soliton-like solutions were first obtained for the self-focusing of beams in nonlinear optical media. The process is described by a nonlinear parabolic equation, the solutions of which describe a self-supporting waveguide (so-called *spatial* solitons) (*Akhmanov et al.* 1967).

Ostrovsky (1963) predicted shock waves and later soliton-envelope waves in an unbounded medium (*Ostrovsky*, 1966). At the same time other advances were also made, in particular by *Talanov* (1964). Solitons in resonant media were studied by *McCall* and *Hahn* (1967). An important step accelerating further research was made by *Zakharov* and *Shabat* (1971). They showed that the equation which describes the evolution of the envelope of the electric field amplitude in a medium with cubic nonlinearity (the nonlinear Schrödinger equation (NLS)) belongs to the class of fully integrable nonlinear evolution equation which can be solved by application of the inverse scattering transform (IST). This work, together with that of *Gardner* et al. (1967) for the Korteweg-de Vries (KdV) equation are important milestones in the development of soliton theory. After the publication of these ideas it became apparent that there exists a wide class of nonlinear evolution equations which are completely integrable. These early works initiated an intense development of the theory of optical solitons which continues to the present day. Later, *Hasegawa* and *Tappert* (1973) showed that optical fibres could also sustain envelope solitons, described by a one-dimensional nonlinear Schrödinger equation. The envelope solitons were experimentally discovered by *Mollenauer* et al. (1980).

Nonlinear optical systems are of interest from the standpoint of general soliton theory. The great variety of optical solitons – dynamic and topological – is due to the many different properties of the media involved. These properties include nonlinearity, material and geometric dispersion and the fact that media can be either passive or active. The dynamic soliton is a localized wave excitation that may be reduced to zero by a continuous deformation. An example is the Korteweg-de Vries soliton, or the NLS bright soliton. The topological soliton represents a transient region where the pulse connects two different states of the host medium. Examples include the sine-Gordon (SG) soliton when describing self-induced transparency, and the dark soliton in an optical fibre (see Chaps. 2 and 5).

There are two main categories in the study of optical solitons:

1) Optical solitons in nonresonant media. The studies are related to short pulse propagation in optical fibres with Kerr-type nonlinearity. These have been reviewed by *Vysloukh* (1982), *Dianov* and *Prokhorov* (1986), *Akhmanov* et al. (1986), *Sysakyan* and *Schwartzburg* 1984), *Mollenauer* and *Stolen* (1982), *Abdullaev* et al. (1987), *Mollenauer* (1988), *Hasegawa* (1989) and *Manykin* et al. (1986);

2) Optical solitons in resonant media. Though solitonic regimes in resonant media were predicted by McCall and Hahn as early as 1967, results describing three- and N-wave interactions are more recent: for example *Bolshov* and *Likhansky* (1985), *Radjarman* (1985), *Rupasov* (1984). New problems have arisen from investigations on superfluorescence (*Zakharov et al.*, 1984).

What is the reason for the successful application of soliton concepts to nonlinear optics? Before answering this question, let us recall that the early development of nonlinear optics was based on exploiting the idea of weakly interacting quasi-harmonics, where it was assumed that the anharmonicities were weak and could therefore be treated perturbatively. That approach allowed the prediction of a number of important phenomena such as the generation of the second and higher harmonics, the generation of difference frequencies, four-photon parametric interaction, etc. However, in nonlinear optics many situations exist where the nonlinearity cannot be considered to be a weak perturbation. Then it is sometimes appropriate to use the concept of solitons which is the analog of quasi-harmonics for weakly nonlinear systems. The development of soliton theory revealed that solitons and their sets along with the radiation component represent the final stage in the evolution of an initial wave field for a broad class of nonlinear media.

In this monograph, we present the description of the current state of optical soliton theory and its applications. Chapter 2 deals with the dynamics of optical solitons in fibres. It is shown that the evolution of intensive electromagnetic pulses in optical fibres can be described in terms of a unified nonlinear wave equation, which is a generalized nonlinear Schrödinger (NLS) equation. This equation is then applied to the analysis of different processes in optical fibres, such as self-phase modulation, soliton propagation, and the behaviour of electromagnetic shock waves in waveguides. Experiments on compression and broadening of short, intensive pulses and on the observation of optical solitons and their complexes are discussed. An analysis based on the NLS equation has shown that the joint action of a nonlinearity and anomalous dispersion on the modulated plane wave leads to the exponential growth in the modulation amplitude (the modulational instability (MI)). In this chapter the theory of MI is also developed.

In a real fibre, effects other than nonlinearity and anomalous dispersion may exist, such as losses, higher order dispersion, etc.. These effects influence the soliton by changing its width, velocity, and phase. Such changes are important when considering different applications of optical solitons, in particular, in applications for optical communication systems. Next we discuss different amplification methods for optical solitons: especially, periodic amplification by optoelectronic devices and Raman amplification (*Vysloukh* and *Serkin*, 1985; *Hasegawa*, 1984). The Raman effect is also responsible for a continuous shift of the frequency spectrum of the soliton to lower frequency (*Mitschke* and *Mollenauer*, 1986; *Gordon*, 1986); this is called the soliton self-frequency shift. This effect and the possible existence of multicomponent solitons in fibres are discussed at the end of the chapter.

In Chapter 3 the interaction between optical solitons in waveguides is discussed. This process has relevance to studies on the possible application of the optical solitons to information systems, since the interaction between solitons decreases the rate of information transmission. First, the methods of analyzing soliton interactions in integrable and nonintegrable systems are outlined. These advanced methods have been applied to the analysis of the elastic interaction of two solitons in an ideal fibre. Experimental work on the observation of soliton interactions in an optical fibre (*Mollenauer* and *Smith*, 1988) is discussed. Inelastic soliton interactions due to the presence of loss, Raman scattering, and higher order dispersion in the fibre are studied. The development of new methods describing information transmission over nonlinear channels demands the study of quasi-periodic (finite-gap) states in optical waveguides. For this reason, the interaction between a soliton and a nonlinear periodic wave is investigated. Interesting applications arise from soliton interactions in two tunnel-coupled optical waveguides (the soliton coupler); this problem is considered at the end of Chapter 3.

Chapter 4 considers the statistical dynamics of optical solitons. This is neccessitated by the fact that the source field is only partially coherent, i.e., it has a noise component. In addition, the parameters of optical waveguides can be changed randomly along the direction of pulse propagation, leading to a random change of the soliton parameters. First, solutions of several problems in random pulse dynamics in a presoliton region are discussed. In the near-field an analytical solution is found by making use of path integrals, and in the far-field by the methods of IST and stochastic process theory. The evolution of random signals in the near-soliton region is also discussed.

The evolution of random fields in nonlinear optical systems described by the NLS and SG equations is then analysed. Stochastic soliton dynamics arising from random medium inhomogeneities is studied. Adiabatic dynamics of optical solitons is investigated and spectral features of the waves radiated by optical solitons are evaluated.

Random soliton dynamics in nonlinear waveguides may also arise from a mechanism called "dynamic stochasticity". The relevant problems in the theory of nonlinear oscillators have been thoroughly studied (*Chirikov*, 1979), whereas understanding of systems described by nonlinear partial differential equations is still in its infance. In Chapter 4, we analyse the chaotic dynamics of optical solitons in media with periodic modulated boundaries, and with periodically modulated internal parameters. The mechanisms of dynamic stochasticity of optical solitons in resonators are examined.

In Chapter 5 solitons in resonant media are considered. Optical solitons in systems to which the IST can be applied have been reviewed by *Bolshov* and *Likhansky* (1985). The application of the IST to real systems requires the analysis of nearly integrable nonlinear wave equations. It is therefore necessary to develop a perturbation theory based on the IST. This chapter describes the derivation of soliton solutions for the Maxwell-Bloch equations and for a set of Stimulated Raman Scattering (SRS) equations. With the aid of perturbation theory, the evolution of the SRS solitons under the action of molecular relaxation is examined.

We then present a theory of superfluoresence based on the IST. Having constructed it, *Zakharov, Gabitov* and *Mikhailov* (1984) succeeded in solving a mixed problem for the Maxwell-Bloch equations; this was the first success in application of the IST to mixed problems. We then discuss soliton propagation in media with Kerr nonlinearities and resonance impurities. Analysis has shown that there may exist "autosolitons", where the energy-loss rate balances energy input into the system.

Chapter 6 gives an account of the quantum IST and its application to quantum optics (*Rupasov*, 1982). The appendices include brief summaries of optical waveguides and the theories of the NLS and SG equations.

2. Short Pulses in Optical Fibres

The possibility of maintaining high-energy densities in fibre waveguides over long distances enables us to observe nonlinear optical effects, even for an input intensity as low as 100 mW. This is considerably lower than the threshold for unbounded media. Currently, much literature is available on investigations of nonlinear phenomena in fibres, such as Brillouin scattering (BS) (*Cotter*, 1983; *Stolen*, 1979), stimulated Raman scattering (SRS) (*Grudinin et al.*, 1981; *Dianov et al.*, 1979, 1983), stimulated four-photon processes (*Stolen* and *Leibolt*, 1976; *Kitayama* et al., 1982; *Dianov et al.*, 1983), and harmonic generation (*Gambriquas*, 1983; *Stolen*, 1981; *Aueung* and *Yarive*, 1978; *Prokhorov*, 1983; *Dianov* et al., 1984; *Demchuk* et al., 1984). The single-mode, multi-mode, gradient and step fibres have been studied.

The capacity of fibres acting as transmission media in long-distance communication systems makes investigations of both linear and nonlinear optical phenomena not only scientifically significant, but of great practical value. Optical communication systems (OCS) have a number of advantages over cable systems. These include higher transmittance and noise immunity, small size and mass, and lower cost. However, OCS are associated with difficulties due to broadening of the initial short pulse during its travel along the fibre. This fact deteriorates the OCS transmittance and should be studied separately. The pulse broadening is caused by dispersion, of which three kinds are known:

a) material dispersion, associated with the dependence of material index on frequency;

b) waveguide (inner-mode) dispersion, due to the frequency dependence of the mode propagation constant; and

c) intermode dispersion which is due to the difference between the group velocities of various modes.

There are different ways of eliminating dispersive pulse broadening, for example, by employing gradient fibres where optical paths of different modes are correlated so that the propagation times become equal, or by using single-mode fibres, which lack intermode dispersion, and where material and inner-mode dispersions are small or mutually compensated. The mechanism of parametric frequency transformation that could be used to compensate dispersive distortion of the pulse has also been proposed (*Kravtsov* and *Minchenko*, 1984). Other ways of compensation, together with references on this subject are given by *Lam* and *Garside* (1982). However, one fails to entirely eliminate the effect of fibre dispersion on the pulse in linear systems; this is a serious obstacle to increasing the information transmission rate. Nonlinear pulse propagation regimes in a fibre provide new ways of solving this problem. Under certain conditions, dispersive pulse broadening can be compensated by nonlinear compression that results in a stable nonlinear optical pulse: the soliton.

The use of optical solitons opens new prospects for noninterference transmission, since solitons are highly stable with respect to perturbations caused by fibre nonuniformities, and to external interference. Apart from noise immunity, soliton regimes allow rapid information transmission, some 100 times higher than that in the best linear systems.

In this chapter we will confine ourselves to the consideration of the soliton regimes of optical pulse propagation in fibres, noting that single-mode fibres are most promising in terms of their application to communication systems.

2.1 Solitons in Single-Mode Fibres

Hasegawa and *Tappert* (1973) were the first to point out the possibility of creating a propagating soliton in an optical fibre. They showed that, under certain conditions, the equation describing the evolution of the high-frequency signal envelope is reduced to a nonlinear Schrödinger (NLS) equation. The NLS equation is known to belong to a class of equations which are integrable by the inverse scattering transform (IST) and, hence, to have a soliton solution (*Zakharov* et al., 1980; see also Appendix B). Thus, it was shown that short pulses ($\tau \sim 1$ ps) propagating in a glass fibre in the region of negative group dispersion are in an equilibrium state where pulse broadening due to dispersion and compression due to nonlinear effects are balanced. As a result, a stable pulse, i.e., a soliton is formed. Experimentally, solitons in glass fibres were first observed by *Mollenauer* et al. (1980). They discovered that a pulse width of $\tau \sim 7$ ps and peak intensity of 1.2 W at $\lambda = 1.55$ μm can travel in a quartz single-mode fibre a distance of 700 m with practically no change.

Subsequently, many papers extending earlier investigations appeared (*Bendow* et al., 1980, *Hasegawa* and *Kodama*, 1981; *Dianov* et al., 1983; *Yervick* and *Hermansson*, 1983; *Blow* et al., 1983; *Anderson*, 1983; *Vysloukh*, 1982; *Potasek*, 1984; *Vysloukh* and *Matveeva*, 1985; *Agrawal* and *Potasek*, 1986). These papers treat experimental improvements for generating solitons in single-mode, multimode, and gradient fibres, as well as the effects of energy loss and higher order dispersion. At present, the investigation of soliton propagation in fibres has become an independent branch of nonlinear optics.

We now derive an equation for the envelope of a pulse propagating in a fibre. We take into account the Kerr nonlinearity and consider the conditions under which this equation has a soliton-type solution.

First, we consider a circular dielectric waveguide with core radius $\varrho = \varrho_0$. Chosing the axis z to be along the fibre axis the nonlinear wave equation is

$$\Delta E - \frac{1}{c^2} \left(\frac{\partial^2 D^L}{\partial t^2} \right) = \frac{1}{c^2} \left(\frac{\partial^2 D^{NL}}{\partial t^2} \right) , \qquad (2.1.1)$$

where

$$D^L(t) = \int_0^\infty \varepsilon(t_1) E(t - t_1) dt_1 \qquad (2.1.2)$$

is the linear part of the induction, and

$$D^{NL}(t) = \varepsilon_2 |E|^2 E \qquad (2.1.3)$$

is the nonlinear part.

Note that the expressions given for D^L and D^{NL} are valid for isotropic media, where the permittivity is a scalar function. A solution of (2.1.1) (in the fibre core region) is sought in the form

$$E = e \, \mathrm{Re} \left\{ R(\varrho) A(z, t) e^{igz - i\omega t} \right\} . \qquad (2.1.4)$$

Here, e is a unit vector in the direction of wave polarization; $R(\varrho)$ describes the transverse field modes, where ϱ is a two-dimensional vector in the xy plane; $A(z, t)$ is a slowly varying amplitude:

$$\begin{aligned}
\omega^{-2} \frac{\partial^2 A}{\partial t^2} &\ll \omega^{-1} \frac{\partial A}{\partial t} \ll A , \\
g^{-2} \frac{\partial^2 A}{\partial z^2} &\ll g^{-1} \frac{\partial A}{\partial z} \ll A .
\end{aligned} \qquad (2.1.5)$$

To simplify notations, a derivative will be indicated by a subscript, e.g., $\partial A / \partial z = A_z$. Assuming weak linear dispersion, so that $\varepsilon(t_1)$ is a sharply peaked function, $A(t - t_1)$ can be expanded for small times t_1 into series keeping the terms of the third order:

$$A(t - t_1) = A(t) - t_1 A_t(t) + \frac{1}{2} t_1^2 A_{tt}(t) - \frac{1}{6} t_1^3 A_{ttt}(t) . \qquad (2.1.6)$$

Inserting this into (2.1.2) and using

$$\varepsilon(\omega) = \int_0^\infty \varepsilon(t) \, e^{i\omega t} dt ,$$

$$\varepsilon_\omega = \frac{\partial \varepsilon}{\partial \omega} = i \int_0^\infty t \varepsilon(t) \, e^{i\omega t} dt , \quad \text{etc.,}$$

we obtain

$$\begin{aligned}
D^L = &e R(\varrho) \, e^{igz - i\omega t} \left[\varepsilon(\omega) A(t, z) + i \varepsilon_\omega(\omega) A_t(t, z) \right. \\
&\left. - \frac{1}{2} \varepsilon_{\omega\omega}(\omega) A_{tt}(t, z) - \frac{i}{6} \varepsilon_{\omega\omega\omega}(\omega) A_{ttt}(t, z) \right] .
\end{aligned} \qquad (2.1.7)$$

Since (2.1.1) includes the second derivative with respect to time, we require

$$\frac{1}{c^2}\, D_{tt}^L(z,t) = e R(\varrho)\, e^{igz-i\omega t} \left[-\frac{\omega^2}{c^2} \varepsilon(\omega) A(z,t) \right.$$
$$\left. - \frac{i\omega}{c^2}\, (2\varepsilon + \omega \varepsilon_\omega(\omega))\, A_t(z,t) \right] + a(\omega) A_{tt}(z,t) + ib(\omega) A_{ttt}(z,t) ,$$

where

$$a(\omega) = \frac{1}{c^2} \left[\varepsilon(\omega) + 2\omega \varepsilon_\omega(\omega) + \frac{\omega^2}{2} \varepsilon_{\omega\omega}(\omega) \right] ,$$

$$b(\omega) = \frac{1}{c^2} \left[\varepsilon_\omega(\omega) + \omega \varepsilon_{\omega\omega}(\omega) + \frac{\omega^2}{6} \varepsilon_{\omega\omega\omega}(\omega) \right] .$$

(2.1.8)

Differentiating (2.1.3) twice with respect to time and restricting to terms in the first derivative on the righthand side gives

$$\frac{1}{c^2}\, D_{tt}^{NL} = -e\, \frac{\varepsilon^2}{c^2} |R|^2 R\, e^{igz-i\omega t} \left[\omega^2 |A|^2 A + 2i\omega \left(|A|^2 A \right)_t \right] .$$

(2.1.9)

Then, making use of (2.1.4, 5), the relation

$$\Delta E = \Delta_\perp E + E_{zz} = e\, e^{igz-i\omega t} \left[A\Delta_\perp R + R A_{zz} + 2iRg A_z - g^2 RA \right]$$

(2.1.10)

is obtained. Substituting (2.1.8–10) into (2.1.1) gives

$$A\Delta_\perp R + R A_{zz} + 2igR A_z - g^2 AR + \frac{\varepsilon\omega^2}{c^2}\, AR + i\, \frac{\omega}{c^2}(2\varepsilon + \omega\varepsilon_{\omega\omega}) RA_t$$
$$- a(\omega) R A_{tt} - ib(\omega) R A_{ttt}$$
$$= -\frac{\omega^2}{c^2}\, \varepsilon_2 |R|^2 R |A|^2 A + \frac{2i\omega}{c^2}\, \varepsilon_2 |R|^2 R \left(|A|^2 A \right)_t .$$

(2.1.11)

Assuming that $R(\varrho)$ satisfies the linear equation

$$\Delta_\perp R + \left(\varepsilon k_0^2 - g^2 \right) R = 0 ,$$

$$k_0^2 = \frac{\omega^2}{c^2}$$

(2.1.12)

for intensities far from the self-focusing threshold, then, averaging (2.1.11) over the fibre cross-section gives

$$iA_z + i\, \frac{\omega}{2gc}(2\varepsilon + \omega\varepsilon_\omega) A_t - \frac{a(\omega)}{2g}\, A_{tt} + \frac{1}{2g}\, A_{zz} - \frac{ib(\omega)}{2g}\, A_{ttt}$$
$$= -\frac{\omega^2}{2gc^2}\, \alpha_0\varepsilon_2 |A|^2 A - \frac{i\omega}{gc^2}\, \alpha_0\varepsilon_2 \left(|A|^2 A \right)_t ,$$

(2.1.13)

where

$$\alpha_0 = \frac{\int |R|^4 d\varrho}{\int |R|^2 d\varrho} .$$

(2.1.14)

In glass fibres, the difference between core and cladding refractive indices is usually small ($n_1 - n_2 \ll n_1$). For such fibres an analysis can be carried out analytically. To a good approximation $R(\varrho)$ for a single-mode fibre is Gaussian, $\sim \exp\left[-\varrho^2/2\varrho_0^2\right]$ (*Snyder* and *Love*, 1981). In this case $\alpha_0 \approx 0.5$.

For the HE_{11} mode $g \approx k = \omega\sqrt{\varepsilon}/c$. Then,

$$\frac{\omega}{2c^2g}(2\varepsilon + \omega\varepsilon_\omega) = k' = \frac{1}{v} \ ,$$

$$a(\omega) = kk'' + (k')^2 \ , \tag{2.1.15}$$

$$b(\omega) = k'k'' + \frac{kk'''}{3} \ ,$$

where the primes indicate differentiation with respect to ω. Making use of (2.1.15), and the fact that, to the required accuracy:

$$A_{zz} - \left(\frac{1}{v}\right)^2 A_{tt} \approx -\frac{2}{v}\frac{\partial}{\partial t}\left(A_z + \frac{1}{v}A_t\right)$$

$$\approx i\frac{k''}{v}A_{ttt} + \frac{i\omega^2\varepsilon_2\alpha_0}{vgc^2}\left(|A|^2A\right)_t \ ,$$

we obtain

$$i\left(A_z + \frac{1}{v}A_t\right) - \frac{1}{2}k''A_{tt} - \frac{i}{6}k'''A_{ttt}$$

$$+ \frac{k\varepsilon_2\alpha_0}{2\varepsilon}|A|^2A + i\frac{3\varepsilon_2\alpha_0}{2v\varepsilon}\left(|A|^2A\right)_t = 0 \ . \tag{2.1.16}$$

It is convenient now to transform this result to a reference frame moving with the group velocity, and to introduce dimensionless variables

$$q = \frac{A}{|A_0|} \ , \qquad \tau = \frac{t - z/v}{t_0} \ ,$$

$$\xi = \frac{z}{z_{nl}} \ ,$$

$$z_{nl}^{-1} = \frac{k\varepsilon_2\alpha_0|A_0|^2}{2\varepsilon} \ ,$$

$$t_0 = \left(z_{nl}\left(-k''\right)\right)^{1/2} \ , \tag{2.1.17}$$

$$\alpha = \frac{3}{kvt_0} \ ,$$

$$\beta = -\frac{k'''}{6k''t_0} \ ,$$

where z_{nl} characterizes the nonlinear properties of the fibre (Sect. 2.2.) and $|A_0|$ is a measure of the maximum amplitude of the input pulse.

Now, (2.1.16) takes the form

$$iq_\xi + \frac{1}{2}\, q_{\tau\tau} + |q|^2 q = i\beta q_{\tau\tau\tau} - i\alpha \left(|q|^2 q\right)_\tau \,. \tag{2.1.18}$$

Omitting the terms on the righthand side, this reduces to the nonlinear Schrödinger equation. It has the solution (Appendix B):

$$q_s = q \,\text{sech}\, q_0(\tau - \zeta)\exp\left(\frac{i}{2}\, q_0^2 \xi + i\delta\right)\,, \tag{2.1.19}$$

where q_0, ξ, δ, ζ, are dimensionless parameters defining the soliton amplitude, velocity, phase, and position of the centre, respectively.

Equation (2.1.18) describes the dynamics of the pulse envelope, allowing the effects of self-phase modulation and higher order dispersion. Depending on the spectral region, one or another of these terms dominates the evolution of power and width of the optical pulse. In the following, we will consider various limiting situations when different terms in (2.1.18) are omitted. Note that (2.1.18) takes no account of pulse damping. This may be done by inserting the term $i\Gamma q$, $\Gamma = \gamma z_{nl}$, where γ is a phenomenological parameter equal to the inverse length at which the field amplitude decreases by a factor of e.

Later, along with the nonlinear permittivity, we will use a nonlinear refractive index $n = n_0 + n_2 |E|^2$, where $n_0 = \sqrt{\varepsilon}$, $n_2 = \varepsilon(2n_0)^{-1}$. For SiO_2 (*Tsoar* and *Gerstein*, 1976), $n_2 = 1.2 \cdot 10^{-22}(m/V)^2 = 1.08 \cdot 10^{-13}CGS$. In this case $z_0 = k_0 \alpha_0 n_2 |A_0|^2$ and for $|A_0| = 2 \cdot 10^6$ V/m, $\lambda_0 = 2\pi/k_0 = 1.5$ μm, $k'' = -10^{-26}\sec^2/m$, we have $\xi = 1$ at $z \approx 1$ km and $t_0 \approx 3$ ps.

Note that the dispersive properties of a fibre are sometimes described by the parameter $D(\lambda) = 2\pi c k''/\lambda^2$ or by the dimensionless parameter $\tilde{D}(\lambda) = \lambda^2 d^2 n/d\lambda^2 = 2\pi c^2 k''/\lambda$.

An alternative normalization variables is

$$\tau = (t - z/v)/\tau_0\,, \quad z = z/z_d\,, \quad q = \sqrt{m}A/|A_0|\,.$$

Here, $z_d = \tau_0^2/|k''|$ is the dispersion length, i.e., the length required in the linear approximation for the pulse width to double, τ_0 is the initial pulse duration, and $m = z_d/z_{nl}$. At $z_d = z_{nl}$ the dispersive broadening of the pulse is compensated by nonlinear compression (Sect. 2.2). This balance condition enables us to estimate the critical peak power P_0 for soliton generation

$$\frac{\tau_0^2}{|k''|} = \left(k_0 \alpha_0 n_2 |A_0|^2\right)^{-1}\,.$$

From this equality, we obtain

$$P_0 = S_{eff} I_0 = \frac{S_{eff}cn_0|A_0|^2}{8\pi} = \frac{S_{eff}cn_0|k''|}{8\pi\tau_0^2 k_0 \alpha_0 n_2} \approx 1W\,.$$

Here we took $\tau_0 = 7$ ps, $S_{eff} = 10^{-6}$ cm^2, $\alpha_0 = 0.5$, where $S_{eff} = \pi \int d\varrho \varrho^2 R^2(\varrho)/\int d\varrho R^2(\varrho)$ is the effective cross-section, and I_0 is the critical intensity.

2.2 Self-Phase Modulation of Pulses

One of the effects caused by the dependence of the refractive index on the electromagnetic wave intensity is self-phase modulation (SPM). This effect is noticeable when pulses propagate through sufficiently long fibres. In the region of positive group velocity dispersion ($k'' > 0$), pulse broadening will arise from both self-phase modulation and group velocity dispersion.

For self-phase modulation in fibres we require equation (2.1.13) to be (keeping only first-derivative and a nonlinear terms) in the following form

$$A_z + \frac{1}{v}\, A_t - if|A|^2 A = 0\,, \tag{2.2.1}$$

where

$$f = \frac{\alpha_0 k \varepsilon_2}{2\varepsilon} = \left(z_{nl}|A_0|^2\right)^{-1}\,.$$

The solution of this equation is

$$A(z,\tau) = A(\tau)\, e^{i\phi(z,\tau)}\,, \tag{2.2.2}$$

where $\tau = t - (z/v)$, $\phi = f|A|^2 z$, and $\phi(z = 0, \tau)$ is zero.

The pulse envelope is not affected, but the phase depends on the distance covered and on $|A|^2$, i.e., SMP emerges.

The field in the fibre has the form

$$E(r,t) = eR(\varrho)A(\tau)\, e^{igz - i\omega t + i\phi} \tag{2.2.3}$$

and the pulse frequency is

$$\omega = \omega_0 - \frac{\partial \phi}{\partial t} = \omega_0 - fz\, \frac{\partial |A|^2}{\partial t}\,. \tag{2.2.4}$$

For the Gaussian pulse $|A|^2 = A_0^2 \exp(-\tau^2/\tau_0^2)$ the frequency change $\Delta\omega = \omega - \omega_0$ is

$$\Delta\omega = \frac{2z}{z_{nl}\tau_0^2}\, \tau \exp\left(-\frac{\tau^2}{\tau_0^2}\right)\,. \tag{2.2.5}$$

A plot of $\bar{\Delta\bar\omega} = (\Delta\omega\tau_0 z_{nl}/2z)$ is depicted in Fig. 2.1. A similar plot is obtained for the pulse $|A| = A_0 \operatorname{sech}(\tau/\tau_0)$.

It follows from (2.2.4,5) that the frequency of the leading edge of the pulse is decreased. This results in pulse broadening in the region of positive group dispersion $k'' > 0$ and to pulse narrowing when $k'' < 0$.

A pulse spectrum at the fibre output, defined by the Fourier transform of (2.2.3), is seen to be dependent on the fibre length and output pulse power.

Self-modulation (in the absence of self-focusing effects) was observed for fairly low energy densities in capillary CS_2 filled fibres (*Ippen* et al., 1974).

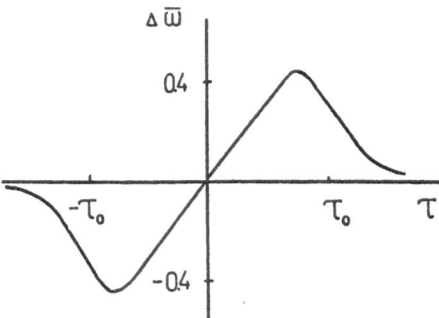

Fig. 2.1. Variation of carrier frequency along a pulse

The analysis of the frequency spectrum was hindered by the use of a multi-mode fibre and a picosecond pulse ($\Delta t = 2.5$ ps), whose width was of the order of the relaxation time of the nonlinearity parameter in CS_2. These drawbacks were eliminated by *Stolen* and *Lin* (1978), who measured spectral broadening of Ar mode-synchronized ($\lambda = 5415$ Å) laser pulses at the single-mode glass fibre output. The measurements were carried out with two fibres. The first had a length of 116 m, a core diameter $d = 4.5$ μm, and 32 db/km loss at the wavelength $\lambda = 5415$ Å. The second fibre was 99 m, had $d = 3.5$ μm, and loss ~ 17 db/km. The frequency radiation spectrum was recorded by a scanning Fabry-Perot interferometer. Such measurements for different peak input powers showed good agreement with theoretical calculations for a Gaussian pulse. Figure 2.2 shows calculated frequency spectra for Gaussian pulses with different peak powers (*Stolen* and *Lin*, 1978). The value of $\phi_0 = L/z_{nl}$ labels each plot.

As mentioned before, the common action of SPM and positive group velocity dispersion (i.e., $k'' > 0$) leads to pulse broadening in long waveguides. Obviously, this will affect information transmission in optical fibre communication systems. A critical length L_c can be defined by the criterion that the high-frequency part of the pulse should be retarded behind the low-frequency part by a pulse length,; then

$$2\tau_0 v = \frac{|\Delta v| L_c}{v} . \tag{2.2.6}$$

Here, $\Delta v = -\Delta \omega v^2 k''$, where Δv and $\Delta \omega$ are the velocity and the frequency differences between high- and low-frequency parts of the pulse. Figure 2.1 indicates that $\Delta \omega \cong 2L_c/\tau_0 z_{nl}$, so that (2.2.5) becomes

$$L_c^2 = z_d z_{nl} . \tag{2.2.7}$$

Equivalently,

$$L_c = \left(\frac{\tilde{D}(\lambda) n_2 |A_0|^2}{2} \right)^{-1/2} c\tau_0 ,$$

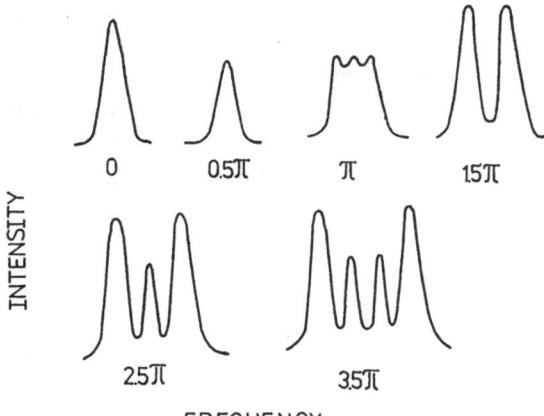

INTENSITY

0   0.5π   π   15π

2.5π   3.5π

FREQUENCY

Fig. 2.2. Calculated spectra of Gaussian pulses at different peak powers. From *Stolen* and *Lin* (1978)

where $\tilde{D}(\lambda) = \lambda^2 d^2 n / d\lambda^2$ and $c\tau_0$ is the pulse length; here $\lambda = 0.5$ μm, $\tilde{D}(\lambda) = 0.08$.

As shown by *Stolen* and *Lin* (1978), the frequency width doubles at $\phi_0 \approx 2$, and for a fibre of 255 m length the corresponding critical power is ~ 180 mW, whereas for Brillouin scattering it is ~ 3 W.

In the case of the negative group dispersion, pulse narrowing occurs. Then (2.2.7) determines the propagation length L where the pulse duration decreases by a factor of two. It is obvious that at $L = z_{nl} = z_d$ the dispersive broadening is completely compensated by the nonlinear compression.

2.3 Soliton Regime of Ultrashort Pulse Propagation. Bright and Dark Solitons

It was shown in Sect. 2.1 that when higher order dispersive and nonlinear effects are neglected, short pulse (SP) propagation in nonlinear optical guides is described by the NLS equation

$$iq_\xi + \frac{1}{2} q_{\tau\tau} + |q|^2 q = 0 . \qquad (2.3.1)$$

Much is known about the conserved densities and the Hamiltonian structure of this equation. These are described (in part) in Appendix B.

To begin, we describe the single soliton solution. A soliton is known to arise when dispersive and nonlinear effects are balanced. Linear dispersion causes different frequency components with different velocities to give rise to wave packet broadening. Nonlinearity leads to a steepening of the wave fronts and, eventually, to the formation of shock waves. Shock waves are predominant in acoustics, where dispersion is usually neglible. When dispersion and nonlinearity are balanced, solitons and nonlinear stationary periodic waves occur. Solitons propagate without change of envelope. Generally speaking, solitons are known to be special solutions of completely integrable equations. A dominant characteristic feature is their elastic collisions with each other. Stationary solutions of nonintegrable nonlinear wave equations (such as the nonlinear Klein-Gordon equation with cubic nonlinearity) are called solitary waves. The collisions of solitary waves are inelastic.

To find the solution of the NLS equation (2.3.1), we take q in the form

$$q(\xi, \tau) = A(\tau - v\xi) \exp(i\phi\xi + i\psi\tau) . \tag{2.3.2}$$

Substituting (2.3.2) into (2.3.1) gives

$$-ivA_\tau - \phi A + \frac{1}{2} A_{\tau\tau} + i\psi A_\tau - \frac{1}{2} \psi^2 A + A^3 = 0 . \tag{2.3.3}$$

The requirement that the terms proportional to A_τ vanish gives the condition $\psi = v$.

Now, we introduce a constant B,

$$\phi + \frac{1}{2} \psi^2 = B , \quad \phi = -\frac{v^2}{2} + B . \tag{2.3.4}$$

Substituting (2.3.4) into (2.3.3) gives

$$\frac{1}{2} A_{\tau\tau} - BA + A^3 = 0 .$$

Multiplying this equation by A_τ and then integrating, we obtain

$$(A_\tau)^2 = P(A) = C - A^4 + 2BA^2 . \tag{2.3.5}$$

For the soliton A, $A_\tau \to 0$ for $|\tau| \to \infty$, and so the integration constant $C = 0$; from (2.3.5) it follows that

$$\int_{A_0}^{A} \frac{dA}{A\sqrt{2B - A^2}} = \pm(\tau - \tau_0) .$$

Evaluating the integral produces

$$A(\tau) = A_0 \operatorname{sech} A_0(\tau - \tau_0) , \quad A_0 = \sqrt{2B} . \tag{2.3.6}$$

Finally, using (2.3.2–4,6), we have the soliton solution

$$q_s(\xi, \tau) = \frac{\sqrt{2B}}{\cosh \sqrt{2B}(\tau - v\xi - \tau_0)} \exp\left(\frac{-iv^2\xi}{2} + iv\tau + iB\xi\right) .$$

Inserting $\sqrt{2B} = 2\eta$ and $v = -2\mu$, we obtain a canonical form of the one-soliton solution

$$q_s(\xi, \tau) = 2\eta \, \text{sech} \, 2\eta(\tau + 2\xi\mu - \tau_0) \exp\left\{ -2i\zeta\tau + 2i(\eta^2 - \mu^2)\xi + i\delta_0 \right\} \quad .(2.3.7)$$

The periodic solution is similarly obtained. This is sought by reversing the elliptic integral in (2.3.5):

$$\int_{A_0}^{A} \frac{dA}{(C - A^4 + 2BA^2)^{1/2}} = \pm(\tau - \tau_0) . \tag{2.3.8}$$

Let us represent $P(A)$ as

$$P(A) = (\alpha_1 - A^2)(A^2 - \alpha_2) ,$$

where

$$\alpha_{1,2} = B \pm \sqrt{B^2 + C} ,$$

and set $A_0 = 0$. Then,

$$\frac{1}{\sqrt{\alpha_1}} F\left(\phi, \frac{\sqrt{\alpha_2}}{\sqrt{\alpha_1}} \right) = \pm(\tau - \tau_0) ,$$

where $\phi = \sin^{-1}(A/\sqrt{\alpha_2})$ and $F(\phi, k)$ is an elliptic integral of the first kind with k being its modulus. Hence,

$$A = \sqrt{\alpha_2} dn \left(\tau\sqrt{\alpha_1}, k \right) , \tag{2.3.9}$$

where $k = \sqrt{1 - k_1^2}$, $k_1^2 = \alpha_2/\alpha_1$, $dn(\cdot)$ is an Jacoby elliptic function. As $C \to 0$ then $\alpha_1 \to 2B$, $\alpha_2 \to 0$, and $k \to 1$, hence we have a periodic lattice of noninteracting solitons. New phenomena are observed for the interaction of solitons with a cnoidal wave. This problem will be considered in detail in Chap. 3.

Note that, unlike a KdV soliton, the amplitude of an NLS soliton is independent of its velocity. The NLS soliton describes a modulated wave packet propagating through a nonlinear dispersive medium with a constant velocity (Fig. 2.3).

The NLS equation is an integrable system (*Zakharov* et al., 1980) having an infinite number of integrals of motion. The first three integrals are often used in the physics of optical solitons. The first integral, the "number of quanta" N, is

$$N = \int_{-\infty}^{\infty} |q|^2 d\tau . \tag{2.3.10}$$

It is readily seen from the soliton solution that $N = 4\eta$. The next integral, the "momentum" P, is

$$P = i \int_{-\infty}^{\infty} \left(q^* q_\tau - q_\tau^* q \right) d\tau . \tag{2.3.11}$$

The third one is the Hamiltonian

$$H = \int_{-\infty}^{\infty} \left(|q_\tau|^2 - |q|^4 \right) d\tau . \tag{2.3.12}$$

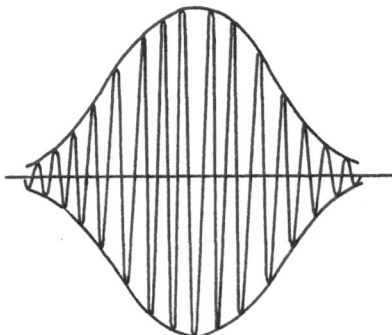

Fig. 2.3. A nonlinear Schrödinger soliton envelope

Let us illustrate the occurrence of an optical soliton in a fibre from the physical point of view. As pointed out in Sect. 2.2, the SPM phenomenon results in a decrease of the pulse frequency in the leading half of the pulse and an increase in frequency in the trailing half. The value of this shift is proportional to the propagation distance down the fibre. When the group velocity dispersion is negative ($k'' < 0$), the group velocity increases with increased frequency. This means that the leading half of the pulse is retarded, whereas the trailing half is advanced, so that the pulse is compressed. At the equilibrium the effects of compression caused by nonlinearity and broadening caused by linear dispersion cancel, resulting in the optical soliton.

The IST allows a detailed study of the evolution of rapidly vanishing initial potentials [i.e., $q(\xi = 0, \tau)$] for the NLS equation. Three types of initial conditions can be distinguished: presoliton, near-soliton, and multi-soliton. Below, we briefly discuss the available analytical results.

2.3.1 Presoliton Region

The nonexistence condition for solitons at the asymptotic limit of large time (i.e., large values of ξ) is

$$\int_{-\infty}^{\infty} |q| \, d\tau \leq \ln(2 + \sqrt{3}) \simeq 1.32 \; .$$

Here, $|q| \to 0$ for $\tau \to \pm\infty$. A general solution to the NLS equation in the presoliton region has the form (*Ablowitz* and *Segur*, 1981)

$$q(\tau, \xi) = \xi^{-1/2} f(\tau/\xi) \exp(i\theta\xi) \; , \tag{2.3.13}$$

where

$$f(\tau/\xi) = -\frac{\ln|a(-\tau/4\xi)|^2}{4\pi} = \frac{1}{4\pi} \ln\left\{ 1 + \frac{|b(\lambda)|^2}{|a(\lambda)|^2} \right\} \Bigg|_{\lambda=-\tau/4\xi} \; .$$

Here a and b are transition matrix coefficients (see Appendix B). The pulse envelope is seen to decrease with its propagation in the fibre according to $1/\sqrt{\xi}$.

It should be noted that for some types of initial conditions in the presoliton region there are complete solutions for the initial value problem. For example:

i) The initial condition is a rectangular pulse

$$q_0(\tau) = \begin{cases} a_0, & 0 \le \tau \le \tau_0 \\ 0, & \tau < 0, \tau > \tau_0 . \end{cases} \tag{2.3.14}$$

Then, for the canonical action, we have (*Zakharov* et al., 1980)

$$n(\lambda) = \frac{1}{\pi} \ln \left(1 - |b(\lambda)|^2\right)^{-1} = \frac{1}{\pi} \ln \left\{ 1 + \left[\sin^2 \left(\tau_0 \sqrt{\lambda^2 + a_0^2} \right) \right] \right.$$

$$\left. \times \left[\frac{\lambda^2}{a_0^2} + \cos^2 \left(\tau_0 \sqrt{\lambda^2 + a_0^2} \right) \right]^{-1} \right\} . \tag{2.3.15}$$

For $a_0 \tau_0 < \pi/2$ there are no solitons.

ii) The initial condition is a periodically modulated hyperbolic secant

$$q_0(\tau) = a_0 \mathrm{sech}(\tau/\tau_0) \exp(id\tau) . \tag{2.3.16}$$

In this case, the variable $n(\lambda)$ is (*Satsuma* and *Yajima*, 1974)

$$n(\lambda) = -\frac{1}{\pi} \ln \left\{ 1 - \mathrm{sech}^2 \left[\pi \tau_0 \left(\lambda - \frac{d}{2} \right) \right] \sin^2(\pi a_0 \tau_0) \right\} . \tag{2.3.17}$$

The condition for the absence of soliton is $a_0 \tau_0 < 1/2$.

2.3.2 Near-Soliton Region

In this case, the initial condition has the form

$$q = q_s + f(\xi = 0, \tau) ,$$

where q_s is the soliton state, and $f(\xi = 0, \tau)$ is an arbitrary rapidly decreasing function. By a proper choice of $f(\xi = 0, \tau)$, one can describe different types of modulation of the soliton state, modulating either the amplitude or the phase. *Satsuma* and *Yajima* (1974) studied the evolution from the initial state

$$q = q_s(1 + \varepsilon) , \qquad q_s = \mathrm{sech}\, \tau .$$

For $\varepsilon < 0$ the process starts with pulse broadening, for $\varepsilon > 0$ from pulse compression. The analysis shows that the interaction between the soliton and the radiation field results in a phase shift of the soliton and the deformation of the radiation field. A numerical analysis shows that a continuous spectrum (radiation field) asymptotically vanishes, and that for $\xi > 0.15$ the soliton component is essentially isolated. The radiation component vanishes like $1/\sqrt{\xi}$, as predicted by theory.

2.3.3 Multi-Soliton Region

Together with the one-soliton state, there exist N-soliton states that are also exact solutions of the NLS equation. They can be constructed with the help of the IST (Appendix B) (*Zakharov* et al., 1980). The initial state for the N-soliton bound state is

$$q(0, \tau) = N \text{ sech } \tau , \qquad (2.3.18)$$

where N is an integer.

For $N = 2$ the solution of (2.3.1) is

$$q_s(\xi, \tau) = 4 \exp \left(\frac{-i\xi}{2} \right) \frac{\cosh(3\tau) + 3 \exp(-4i\xi) \cosh(\tau)}{\cosh(4\tau) + 4 \cosh(2\tau) + 3 \cos(4\xi)} . \qquad (2.3.19)$$

This expression shows that the bound states of two solitons oscillate in ξ, and that the oscillation period is $\pi/2$. So, during propagation in the fibre the pulse shape periodically oscillates.

Figures 2.4 and 2.5 show results of numerical simulation of $N = 2$ and $N = 4$ soliton propagation in an ideal single-mode fibre. Soliton periodicity in ξ is observed. Moreover, for the initial conditions (2.3.18), $q(\xi = 0, \tau = 0) > 1$, and thus, initially the pulses are strongly narrowed. This is utilized in fabricating fibre optic compressors.

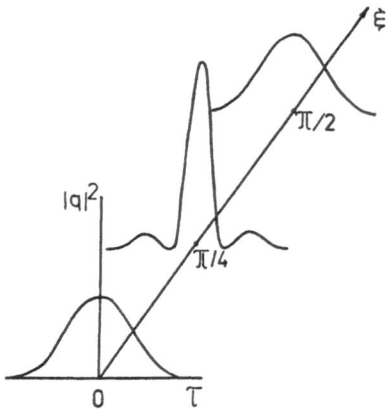

Fig. 2.4. Evolution of the $N = 2$ soliton in an ideal single-mode fibre, numerical simulation

For $q(0, \tau) = (N + \varepsilon) \text{sech } \tau$, $|\varepsilon| < 1/2$, the final stage of the pulse evolution is the N-soliton state. *Satsuma* and *Yajima* (1974) give a general formula to estimate the number of solitons in the asymptotic state evolved from the initial condition $q(0) = a$ sech b. The number of solitons is $n = [a/b + 1/2]$, where $[\dots]$ denotes the integer part of the expression.

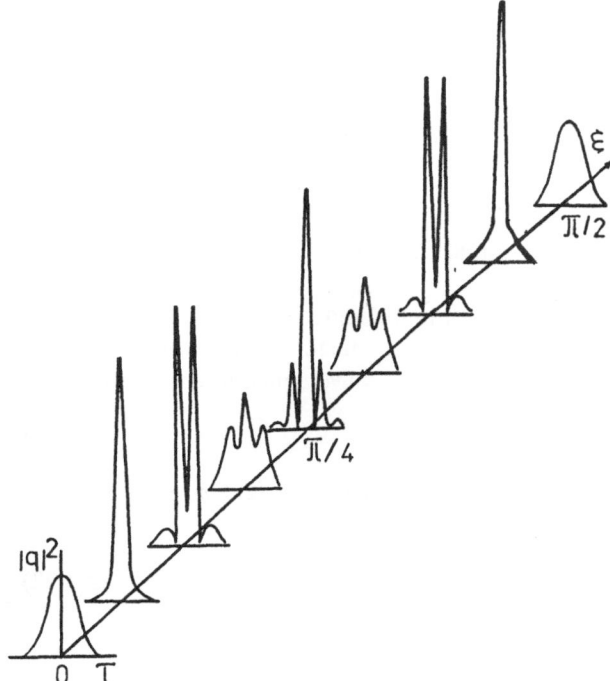

Fig. 2.5. Evolution of the $N = 4$ soliton in a single-mode fibre, numerical simulation

The decay of the smooth "large" initial condition $q(0, \tau)$ into solitons is also of interest. "Large" means that the number of eigenvalues in the Zakharov-Shabat system (B.12) is $n \gg 1$.

Functions $q(0, \tau)$ are assumed to be smooth, namely,

$$\frac{q_\tau(0, \tau)}{q(0, \tau)} \ll 1 ,$$

so that nonlinear geometric optics is applicable to (2.3.1). The relevant linear spectral problem (B.2) can be analyzed by the WKB method. The calculations show (*Zakharov* et al., 1980) that the reflection coefficient is exponentially small, and that the asymptotic state can be approximated by the N-soliton solution. Amplitudes of solitons η_n in this solution are evaluated with the help of the Bohr quantization rule

$$\oint \sqrt{|q(0)|^2 + \eta^2}\, d\tau = 2\pi \left(n + \frac{1}{2} \right) . \tag{2.3.20}$$

It should be noted that the binding energy is zero, i.e., for any small deviation from the exact N-soliton form, the bound state is decomposed into free solitons whose velocities can be different and are proportional to the perturbation.

Burzlaff (1988) and *Kivshar* (1989) derived the number of solitons N to be

$$N = \left[\frac{1}{2} + \frac{F}{\pi} \right] ,$$

where

$$F = \int_{-\infty}^{\infty} |q(0,\tau)| d\tau .$$

Manakov (1973) considered for the initial condition $q(0,\tau) = i\beta \exp(-\alpha\tau^2)$ and found that $N = [1/2 + 2\beta/\alpha\pi]$.

In addition to the solutions which exist in the region of anomalous dispersion $k'' < 0$, there exist other types of soliton states in the region of normal dispersion $k'' > 0$. In the region of positive dispersion the wave equation has the form

$$iq_\xi - \frac{1}{2} q_{\tau\tau} + |q|^2 = 0 . \tag{2.3.21}$$

Applying a similar procedure to that following equation (2.3.2), we obtain the "dark soliton" solution:

$$q_d(\xi,\tau) = \pm q_0 \tanh\left[q_0(\tau - \tau_0 - v\xi) \right] \exp\left\{ \frac{iv^2\xi}{2} + iq_0^2\xi - iv\tau \right\} . \tag{2.3.22}$$

For $v = 0$, we have

$$q_d = \pm q_0 \tanh\left[\eta(\tau - \tau_0) \right] \exp\{iq_0^2\xi\} . \tag{2.3.23}$$

From this, the wave intensity is

$$|q_d|^2 = q_0^2 \left[1 - \text{sech}^2(q_0\tau) \right] . \tag{2.3.24}$$

The solution (2.3.24) describes a stationary "hole" in the intensity of a CW beam or in a sufficiently long pulse; this accounts for the name "dark soliton".

Hasegawa and *Tappert* (1973) found a solution of (2.3.21) in the form

$$q(\tau) = A_0 \left[B^{-2} - \text{sech}^2\tau \right] \exp\left\{ i\xi^2\tau + i\phi(\tau_0) \right\}$$

$$\sin\phi(\tau) = \frac{B\tanh(\tau)}{\left[1 - B^2\text{sech}^2(\tau) \right]^{1/2}} , \quad |B| < 1 .$$

In the limiting case $B = 1$ this coincides with the fundamental dark soliton. For $B < 1$, the pulse may be referred to as "grey" soliton (*Tomlinson* et al., 1989), and B can be considered as a parameter which measures darkness. In contrast to bright solitons, whose phase is constant across the entire pulse, the phase of dark solitons changes by π at the centre. *Christodoulides* and *Joseph* (1984) showed that taking account of a radial dependence alters the asymptotic behaviour of a soliton; it vanishes as $\xi \to \infty$. Recently, *Gredeskul* et al. (1990) studied the evolution of a wide class of initial potentials for equation (2.3.21) using the IST method and *Rothenberg* and *Heinrich* (1992) observed the formation of dark soliton trains in optical fibers.

2.4 Observations of Optical Solitons

Since *Mollenauer* et al. (1980) made the first observations of optical solitons in a fibre, solitons have played an important role in pulse compression. Experimental observations of optical solitons in fibres are now possible for two reasons. First, since 1980, single-mode, fused-quartz optical fibres with anomalous dispersion and low loss (on the order of ~ 0.2 db/km) at a wavelength of $\lambda > 1.3$ μm have been fabricated. Second, tunable, mode-synchronized ($\lambda = 1.5$ μm), colour-centre lasers have also been developed.

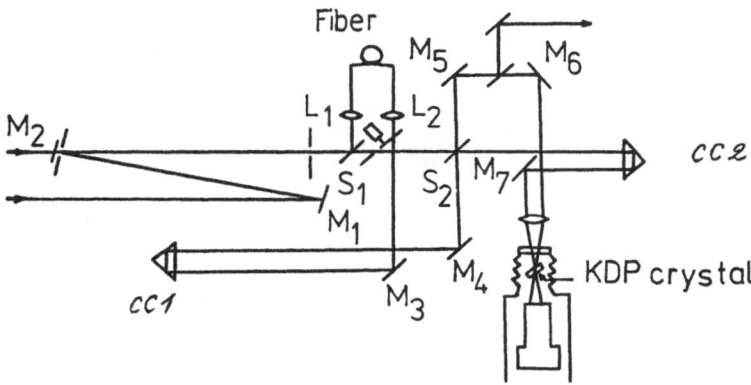

Fig. 2.6. Experimental setup of *Mollenauer* et al. (1980).
M_{1-4} – mirrors, S_1 – beam splitter, the autocorrelator (S_2, M_5, M_6, M_7, CC_2, etc.)

An experimental set-up for observing optical solitons is shown in Fig. 2.6. It comprises a beam former M_1, M_2, A_1, A_2, together with a beam splitter S_1, which splits the beam between the laser and fibre channels. The units S_2, M_5, M_6, M_7, and CC_2 form an autocorrelator.

The radiation source is a colour-center laser; specifically, F_2^+-centres in a NaCl crystal. The tunable range is $\lambda = (1.35-1.75)$ μm. Laser pulses are directed into a single-mode optical fibre. Output pulses are processed with the autocorrelator, where the autocorrelation function of the pulse intensity after traversing the fibre is measured.

The generated pulses must be structureless (smooth) and well-shaped. For this purpose, the resonator lengths of the pump and colour-centre lasers were chosen so that an input pump pulse corresponds to the generated output pulse. The colour-centre laser has a sapphire plate of 4 mm thickness to guarantee that $\Lambda = \Delta f \, \Delta t = 0.18$, indicating the bandwidth of limited pulses. Here, Δf and Δt are pulse half-widths in frequency and time, respectively. The value of Λ should lie between its minimal value for the $\mathrm{sech}^2 \tau$ profile ($\Lambda = 0.315$) and that for a decreasing exponent ($\Lambda = 0.11$). The initial pulse must have a correct shape to observe soliton effects reliably.

The measurement of the autocorrelation intensity function determines the shape of the pulse envelope in the optical fibre. If the envelope is stationary then this implies the existence of solitons. The autocorrelation function was measured using a procedure based on the generation of a second harmonics in a KDP crystal. When the length of the fibre is equal to one-half of the soliton period $z_s = \pi z_d/2$, the largest change in the pulse envelope occurs at the far end of the fibre. If we take $\tau_0 = 7$ ps, $|D| = 2\pi c|k''|/\lambda_0^2 = 15$ ps/nm·km, $\lambda_0 = 1.55$ μm, then $z_s = 675$ m.

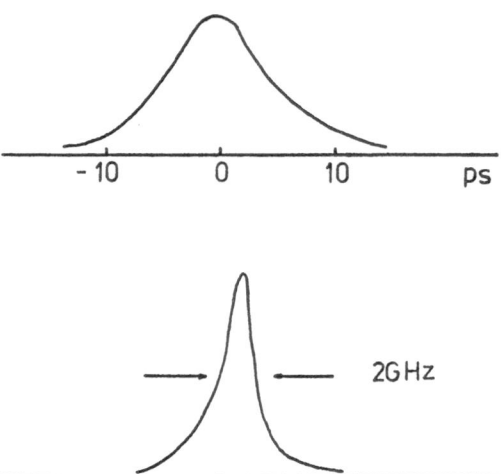

Fig. 2.7. The temporal envelope of a pulse and its frequency spectrum

In the experimental work, short optical pulses with $\tau = 7$ ps and $\lambda = 1.75$ μm were generated at the input to a single-mode fibre with $L = z_s/2$. The shape of a typical pulse, together with its spectrum, are depicted in Fig. 2.7. The input power was varied over the range $P = 0.3$ W to $P = 22.5$ W. The threshold intensity for soliton creation under these conditions is $I_0 = 10^6$ W/cm². With a fibre cross-section $S = 10^{-6}$ cm², the theoretical estimate gives $P_{th} = 1$ W.

The experimental value P_{exp} is 1.2 W. When the input pulse power was varied over the given range, the output pulse varied as depicted in fig. 2.8. With $P_0 = 0.3$ W, the influence of nonlinear effects on the short pulse dynamics is neglible. Only dispersive pulse broadening is observed. When P_0 is increased to 1.2 W, the pulse profile is stationary and consistent with the creation of a single soliton. This is in agreement with the theoretical estimate $P_{th} = 1$ W. For $P_0 = 5$ W pulse compression occurs, and the output trace indicates an $N = 2$ soliton state. For $P_0 = 11.4$ W three peaks appear in the autocorrelation trace,

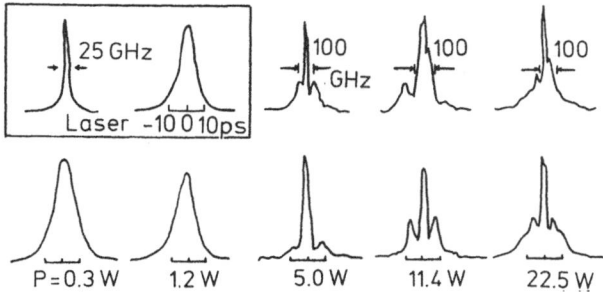

Fig. 2.8. Experimental autocorrelation of intensity at the fibre output vs. the input signal power. Top: the frequency spectra

implying that two peaks are present on the pulse intensity profile, consistent with $N = 3$ state. When P reaches 22.5 W, a four-soliton state is created. According to the NLS equation the critical value of the pulse power should increase according to ratio $1^2 : 2^2 : 3^2 : 4^2$ to allow the generation of higher order solitons. The periodicity of higher-order solitons was also observed by *Mollenauer* and *Stolen* (1982). Measurements were conducted with a single-mode fibre of length $L = z_0 = 1.3$ km. The relevant data is presented in Fig. 2.9, which shows measured pulse shapes and frequency spectra at distance z_0. The pulse shape and spectrum return to their input value at $z = z_0$.

For dark solitons, where $k'' > 0$, experiments have been carried out by *Emplit* et al. (1987) and *Krokel* et al. (1988). They studied the propagation of 0.3 ps pulses in a single-mode fibre within the range of $P \sim 0.4$ to 40 W. Sets of dark solitons were observed whose properties were consistent with numerical simulations. More recently *Rothenberg* and *Heinrich* (1992) observed the formation of dark soliton trains in optical fibers.

2.5 Electromagnetic Shock Waves

It was shown in Sect. 2.1. that the one-dimensional wave equation for a linearly polarized pulse envelope has the form of (2.1.16). Ignoring the third-order dispersion effects, we will inspect this equation in the frequency region wherein the second-order dispersion is zero ($k'' = 0$). In this case, (2.1.16) becomes

$$iq_\xi + |q|^2 q + i \left(|q|^2 q\right)_\tau = 0 , \qquad (2.5.1)$$

where

$$\tau = \frac{vk}{3} \left(t - \frac{z}{v}\right) , \quad \xi = \frac{z}{z_{nl}} .$$

The last term in (2.5.1) is important in the study of short pulse propagation in the fibre over long distances. A solution of (2.5.1) is sought in the form (*Anderson* and *Lisak*, 1983)

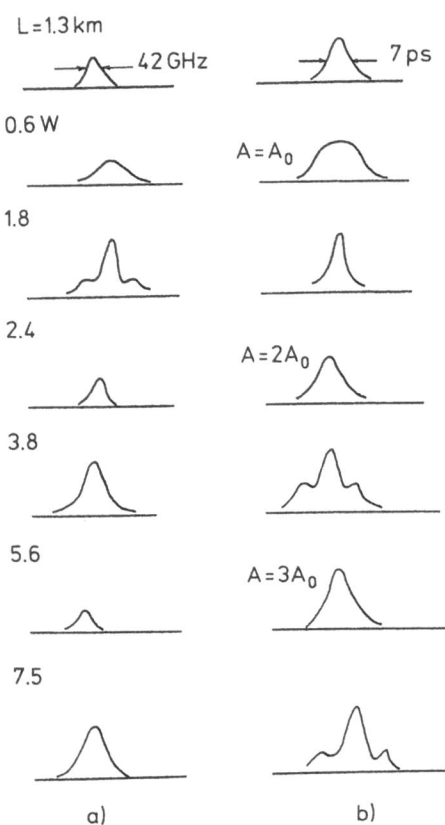

L = 1.3 km

0.6 W

1.8

2.4

3.8

5.6

7.5

a)   b)

Fig. 2.9a,b. Experimental signal output for a fibre with $L = 1.3$km: a) the frequency spectra, b) the autocorrelation shapes

$$q(\xi, \tau) = \varrho \exp(i\phi) .$$
(2.5.2)

Equation (2.5.1) is then replaced by

$$\phi_\xi + \varrho^2 \phi_\tau = \varrho^2 ,$$
(2.5.3)

$$\varrho_\xi + 3\varrho^2 \varrho_\tau = 0 .$$
(2.5.4)

Equation (2.5.4) can be solved in the implicit form

$$\varrho^2 = f(\tau - 3\xi \varrho^2) ,$$
(2.5.5)

where f is an arbitrary function determined by the initial pulse profile. The solution of (2.5.5) may be written in another form

$$\tau = 3\xi \varrho^2 + g(\varrho^2) ,$$
(2.5.6)

where g is the inverse function $g = f^{-1}$. Using the method of characteristics, we also find that

$$\phi = \tau + h\left(3\xi\varrho + \varrho g'(\varrho^2)\right) , \quad g'(\varrho^2) \equiv \frac{dg(\varrho^2)}{d\varrho^2} , \tag{2.5.7}$$

where h is an arbitrary function determined by the initial condition. As seen from (2.5.6,7), the amplitude and frequency are asymmetric due to the terms proportional to ξ.

We now find explicit solutions for two different initial pulse profiles, the Gaussian, and the sech profiles.

1. The Gaussian packet. Let

$$\varrho^2(0, \tau) = \exp\left(-\frac{\tau^2}{\tau_0^2}\right) . \tag{2.5.8}$$

From (2.5.5), we have $f(\tau) = \exp(-\tau^2/\tau_0^2)$, and, thus,

$$g(\varrho^2) = \pm\tau_0 \left(\ln\frac{1}{\varrho^2}\right)^{1/2} .$$

This leads to

$$\varrho^2(\xi, \tau) = \exp\left[-\frac{(\tau - 3\xi\varrho^2)^2}{\tau^2}\right] , \tag{2.5.9}$$

and, thus,

$$\tau = 3\xi\varrho^2 \pm \tau_0 \left(\ln\frac{1}{\varrho^2}\right)^{1/2} , \tag{2.5.10}$$

where the signs (+) and (-) refer to the trailing and leading edges of the pulse, respectively. It follows from (2.5.10) that the local increment in velocity is inversely proportional to the square of the amplitude. This leads to a shift of the pulse peak to the trailing edge, i.e., to an asymmetric change of the pulse shape. The self-steepening of the trailing part of the pulse leads to the appearance of an electromagnetic (EM) shock wave, whose existence was first predicted by *Ostrovsky* (1966). From (2.5.10) it follows that the distance ξ_{cr}, where the derivative ϱ_τ becomes infinite is

$$\xi_{cr} = \left(\frac{e}{2}\right)^{1/2} \frac{\tau_0}{3} . \tag{2.5.11}$$

The critical distance is approximately equal to the traversal length over which the pulse peak is shifted by a distance of the order of the pulse width (Fig. 2.10).

2. The sech-profile. For a profile of the form

$$\varrho^2(0, \tau) = \operatorname{sech}^2\frac{\tau}{\tau_0} , \tag{2.5.12}$$

we have a similar estimate for the critical distance for the development of the shock wave

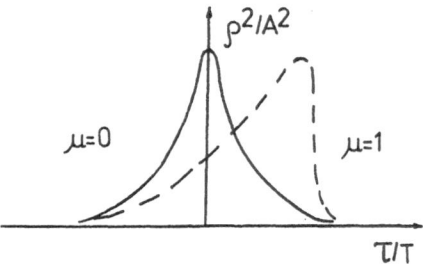

Fig. 2.10. Steepening of the Gaussian profile when propagating in a fibre. The solid curve corresponds to $\mu = 0$, dashed curve to $\mu = 1$. From *Anderson* and *Lisak* (1983)

$$\xi_{cr} = \sqrt{3}\frac{\tau_0}{4} \ . \tag{2.5.13}$$

When considering the propagation of short pulses in real fibres, the effect of damping has to be taken into account. For the linear damping case, (2.5.4) is replaced with

$$\varrho_\xi + 3\varrho^2\varrho_\tau + \Gamma\varrho = 0 \ ,$$

which has the solution

$$\varrho^2 = \exp(-2\Gamma\xi)f\left\{\tau - \frac{3}{2\Gamma}\left[\exp(2\Gamma\xi) - 1\right]\right\} \ . \tag{2.5.14}$$

Repeating the above calculation, the critical distance for a shock wave to appear is then given by

$$\xi_{cr} = -\frac{1}{2\Gamma}\ln\left[1 - (2e)^{1/2}\frac{\Gamma\tau_0}{3}\right] \ . \tag{2.5.15}$$

It follows from (2.5.15) that damping increases the length for shock wave development. However, for $\Gamma > \Gamma_{cr} = 3/\tau_0\sqrt{2e}$, the shock wave does not occur.

We now estimate the distance for the formation of a shock wave. We consider the case of a capillary fibre filled with CS_2. For $\tau_0 = 1$ ps, $n_2 = 10^{-7}$ CGS and $I_0 = 10^{10}$ W/cm², we have $\xi_{cr} = 1$ m. In the glass fibre case where $n_2 = 10^{-13}$ CGS and $I_0 = 10^{11}$ W/cm², $\tau_0 = 10^2$ fs, we get $\xi_{cr} = 20$ cm.

To observe EM shock waves, it is convenient to investigate the nature of the spectral broadening of the pulse. In particular, for $Q \gg 1$, where $Q = n_2\varrho_0^2z/c\tau_0$, the pulse spectrum becomes strongly asymmetric (*Akhmanov* et al., 1986). The reason for this asymmetry lies with the phase term in (2.5.3).

Nesterova and *Aleksandrov* (1985) have attempted to interpret observations of pulse spectrum broadening in capillary CS_2-filled fibres in terms of the formation EM shock waves. One should bear in mind, however, that nonlinearity under these conditions is of a relaxation character.

Such a nonlinear relaxation character in liquids is caused by the time of the nonlinearity transition to a steady-state $\tau_R \sim 10^{-12}$ sec, and its account is already important in picosecond regions. In this case, the equation for nonlinear addition to the refraction index takes the form

$$\tau_R \frac{\partial n}{\partial t} + \Delta n = \frac{1}{2} n_2 |A|^2 .$$

2.6 Influence of Higher Order Dispersion and Dissipation on Soliton Dynamics

As discussed before, the soliton regime for propagation of optical pulses is feasible in single-mode fibres. This is due to the fact that the dynamics of a short-optical pulse envelope in a fibre is described by (2.1.18), which becomes the NLS equation when terms on the righthand side are neglible. However, there are some regimes of pulse propagation where it is necessary to take these terms into account. In particular, to describe effects of dispersive pulse-broadening in the frequency region where k'' is close to 0, one needs to take higher order dispersion into account (in (2.1.18) this is the term in β); taking account of nonlinear dispersion (the term in α) leads to a change in pulse shape and the formation of shock waves, as shown in the previous section. We will consider the influence of these terms on the NLS soliton dynamics using a suitable perturbation theory (*Abdullaev* et al., 1986).

We construct the solution of (2.1.18) using perturbation theory for solitons based on the IST (*Karpman* and *Maslov*, 1977; *Kaup* and *Newell*, 1978; see also Appendix B). The solution is sought in the adiabatic approximation, where it is assumed that the soliton retains its integrity, although its parameters slowly change. Substituting the perturbation terms into (B.25–28), we obtain

$$\frac{dq_0}{d\xi} = 0 , \quad \frac{d\delta}{d\xi} = 0 , \quad \frac{d\zeta}{d\xi} = q_0^2(\alpha - \beta) , \quad \mu = 0 . \tag{2.6.1}$$

These equations can also be derived using conservation laws for the NLS equation. It follows from (2.6.1) and (2.1.19) that the soliton amplitude and phase remain constant, whereas the velocity changes according to the formula

$$v_s = v_g - q_0^2 v_g^2 (\alpha - \beta) \sqrt{-k''/z_{nl}} . \tag{2.6.2}$$

Next, we will find a correction (based on first-order perturbation theory in the adiabatic limit) to the solution

$$q = q_s \left(1 + \omega(\xi, \tau)\right) ,$$

where $\omega(\xi, \tau)$ is the correction from the continuum spectrum of the linear spectral problem.

Making use of the appropriate formulas obtained from perturbation theory (*Karpman* and *Maslov*, 1977), we derive the expression for the soliton envelope

$$q = q_0 \text{sech } q_0 \left[\tau + q_0^2(\beta - \alpha)\xi\right] \exp(2iq_0^2\xi)$$
$$\times \left[1 + i\alpha q_0\tau - 3iq_0\left(\frac{\alpha}{2} + \beta\right)\tanh(q_0\tau)\right] , \tag{2.6.3}$$

where $\delta_0 = \zeta_0 = 0$,

Expression (2.6.3) shows that the perturbative effect of the terms on the right hand side of (2.1.18) is a change of soliton velocity and phase. To estimate the size of the effect, e.g., for a single-mode fibre whose group dispersion vanishes for $\lambda = 1.27$ μm, we will use approximate formula (*Hasegawa* and *Kodama*, 1981):

$$k'' = \frac{\lambda_0^3}{2\pi c^2}\frac{\partial^2 n}{\partial \lambda^2} = \frac{\lambda_0}{2\pi c^2} 5.3 \cdot 10^{-2}\left[\frac{1.27}{\lambda_0} - 1\right]$$
$$\approx 9 \cdot 10^{-28}\left[1.27 - \lambda_0(\mu\text{m})\right]\left[\sec^3/\text{cm}\right] , \tag{2.6.4}$$

$$k''' \approx 2.3 \cdot 10^{-42}\sqrt{\lambda_0(\mu\text{m})}\,(\lambda_0(\mu\text{m}) - 1)\left[\sec^3/\text{cm}\right] .$$

Taking the peak input value for the pulse amplitude to be $A_0 = 10^6$ V/m, the following expression is derived from (2.1.17)

$$\alpha = \frac{3}{2c}\left(\frac{\lambda_0 n_2|A_0|^2}{-k''\pi}\right)^{1/2} = 10^{-4}\left(\frac{\lambda_0}{\lambda_0 - 1.27}\right)^{1/2} , \tag{2.6.5}$$

where $\lambda_0 = 2\pi c/\omega$ is the radiation wavelength in vacuum.

With $\lambda_0 = 1.3$ μm, $\alpha = 0.66 \cdot 10^{-3}$ and $\beta = 0.69 \cdot 10^{-3}$, whereas for $\lambda = 1.5$ μm where damping in the fibre is smallest, $\alpha = 2.6 \cdot 10^{-4}$ and $\beta = 5.4 \cdot 10^{-5}$. For the latter case, we obtain from (2.6.2) with $q_0^2 = 1$ the relation $v_s - v_g = -10^{-7}v_g$. Such a decrease in velocity for a 0.1 ps pulse results in a retardation of approximately 10 pulse-widths over a distance of 2 km.

The resulting change of velocity, as well as the occurrence of frequency modulation, is consistent with numerical calculations reported by *Vysloukh* and *Serkin* (1983) and *Khikmatov* (1987), for the case $\beta = 0$. *Vysloukh* and *Serkin* have studied numerically the influence of nonlinear dispersion on the dynamics of a multi-soliton pulse. The presence of the perturbation is shown to give rise to an unstable bound state of the NLS soliton, with decay caused by the preferential enrichment of the high-frequency part of the spectrum. As a result, an intense short pulse is created in the leading envelope half. It may be used to generate SPs of the order of tens of femtoseconds.

Note that (2.1.18) for the case $\beta = 0$ belongs to the class of fully integrable equations (*Kaup* and *Newell*, 1978). *Golovchenko* et al. (1986) evaluated the first three integrals of motion for this equation and showed, in addition, the feasibility of existing soliton states in the region of positive group dispersion. This is the essential difference between this equation and the NLS equation.

Note that one can construct an exact solution of (2.1.18) when $\beta = 0$, namely (*Okhura* et al., 1987)

$$q = \varrho(z) \exp \{ -i[k\xi - v\tau - \theta(z)] \} , \quad z = \tau - v\xi ,$$

$$k = \frac{1}{2} v^2 - \frac{1}{2}(1 - \alpha v)\varrho_0^2 - \frac{1}{8} \alpha^2 \varrho_0^4 ,$$

$$\varrho(z) = \frac{\varrho_0}{2 - v} \left[\cosh^2(\mu z) + \frac{\nu - 1}{2 - \nu} \right]^{-1/2} ,$$

$$\theta(z) = \frac{3\alpha\varrho_0^2}{2\mu\sqrt{1 - \nu}} \tan^{-1} \left[\sqrt{1 - \nu} \tanh(\mu z) \right] ,$$

$$\mu^2 = (1 - \alpha v)\varrho_0^2 + \alpha^2 \varrho_0^4 ,$$

$$\nu = \frac{(1 - \alpha v)\varrho_0^2}{\mu^2} .$$

The effect of dissipation on soliton dynamics for the case of weak damping can also be taken into account if perturbation theory and conservation laws are applied. Thus, for a soliton

$$q = q_0 \mathrm{sech}\, q_0(\tau - \zeta) \exp \left[i\delta + \frac{1}{2i} \int_0^\xi q_0^2 d\xi \right] , \tag{2.6.6}$$

the energy conservation law leads to the following equation

$$\left(\frac{\partial}{\partial \xi} + 2\Gamma \right) \int_{-\infty}^{\infty} |q|^2 d\tau = 0 , \tag{2.6.7}$$

which implies that the soliton amplitude decreases according to

$$q(\xi) = q_0 \exp(-\Gamma\xi) . \tag{2.6.8}$$

Substitution of this expression into (2.1.19) yields

$$q = q_0 \exp(-2\Gamma\xi)\mathrm{sech} \left[q_0(\tau - \zeta)\exp(-2\Gamma\xi) \right]$$

$$\times \exp \left[i\delta + +\frac{iq_0^2}{8\Gamma}(1 - \exp(-4\Gamma\xi)) \right] . \tag{2.6.9}$$

The use of perturbation theory requires weak damping: for a fibre whose loss is 0.2 dB/km, one obtains $\gamma = 0.023$ km^{-1}, and with $A_0 = 10^6$ V/m, $\lambda = 1.3$ μm, the final result is $\Gamma = \gamma z_{nl} = 0.09$. To decrease the amplitude by a factor of e, the soliton has to travel a distance of the order of 22 km. Note that for $k'' = 0$ the soliton can co-exist with shock waves. *Wai* et al.(1987) showed that fundamental solitons continue to exist all the way down to the zero-dispersion wavelength.

We now consider an investigation of the equation for the pulse envelope at the point of zero-dispersion, where $k''(\omega_0) = 0$,

$$iq_\xi - \frac{iq_{\tau\tau\tau}}{6} + |q|^2 q = 0 \ . \tag{2.6.10}$$

Here,

$$\tau = \frac{t - z/v}{t_0} \ , \quad \xi = \frac{z}{z_{nl}} \ ,$$

$$t_0 = (k''' z_{nl})^{1/3} \ .$$

Dessaix et al. (1990) showed that this equation could be solved using the substitution

$$q(\xi, \tau) = \varphi(\tau, \xi) \exp(-i\Omega\tau) \ . \tag{2.6.11}$$

Neglecting terms of Ω^{-2} order, they found that the solution has the form

$$q = q_0 \mathrm{sech}(\tau - u\xi) \exp\left[-i(\tau - u\xi)\Omega - i\left(\frac{\Omega^3}{3} - \frac{q_0^2}{2}\right)\xi\right] \ ,$$

where $u = \Omega^2/2$ and $q_0^2 = -\Omega$. If the frequency deviation is small, namely, $\omega_0 - \omega \approx 0.01\omega_0$, we find that $\Omega = (\omega - \omega_0)t_0 \approx -5$, where a value for t_0 of 0.5 ps has been used.

2.7 Amplification of Optical Solitons in Fibres

The feasibility of using optical solitons in fibre optic communication lines and information storage and processing systems is limited for real fibres by dissipation. At present, there are several methods for suppressing transmission loss and for increasing information transmission rates by using optical solutions. These methods include the amplification of solitons in regenerators using semiconductor lasers (*Hasegawa*, 1983), or by injecting a plane wave with the same frequency and phase as the soliton (*Hasegawa* and *Kodama*, 1982) or using stimulated Raman scattering (SRS) in fibres (*Vysloukh* and *Serkin*, 1983; *Hasegawa*, 1983). We briefly describe these approaches.

Consider first the injection of a plane wave "pump" together with a soliton into a fibre, where the frequency and phase of the pump wave are the same as the soliton's (*Hasegawa* and *Kodama*, 1982). The initial condition in this case is a soliton in the field of a continuous wave

$$q(0, \tau) = q_s(0, \tau) + f(0, \tau) \ ,$$

where

$$q_s(0,\tau) = 2\eta \ \text{sech}(2\eta\tau) \exp(i\mu\tau + i\delta_0) \ ,$$

$$f(0,\tau) = f_0 \exp(ik\tau + i\theta) \ .$$

Calculations carried out using perturbation theory for the case $f_0 << \eta$ show that the soliton amplitude 2η increases by πf_0 when $k = \mu$ and $\theta = \delta_0$.

An alternative and quite promising approach makes use of energy pumping through the SRS process (*Vysloukh* and *Serkin*, 1983; *Hasegawa*, 1983). The main idea of the method is as follows: a soliton (in the region of anomalous dispersion) and a normally dispersive pump wave propagate together through a single-mode fibre. Under these conditions, the pump wave moves faster than the soliton. Consequently, the soliton propagates in a medium where molecular oscillations are excited and the soliton energy increases due to interaction with these oscillations.

We now proceed to a more detailed description of the interaction between a soliton and a pumping wave (*Dianov* et al., 1985). The process is defined by a set of two coupled equations describing the joint dynamics of complex pumping-wave envelopes $q_p(\xi,\tau)$ and the first Stokes component $q_s(\xi,\tau)$:

$$i\frac{\partial q_s}{\partial \xi} = \frac{1}{2}\frac{\partial^2 q_s}{\partial \tau^2} + |q_s|^2 q_s + R_s|q_p|^2 q_s + iG|q_p|^2 q_s - i\gamma_s q_s \ , \tag{2.7.1}$$

$$i\left(\frac{\partial q_p}{\partial \xi} + \nu\frac{\partial q_p}{\partial \tau}\right) = -\frac{\mu}{2}\frac{\partial^2 q_p}{\partial \tau^2} + R_p|q_p|^2 q_p$$

$$+ R_s|q_s|^2 q_p - iG\frac{\omega_p}{\omega_s}|q_s|^2 q_p - i\gamma_p q_p \ . \tag{2.7.2}$$

These equations are written in a dimensionless form. The dimensionless variables are

$$q_{p,s} = \frac{A_{p,s}}{A_{s0}} \ , \qquad \tau = \frac{t - z/v_g}{\tau_{s0}} \ ,$$

$$\xi = \frac{z}{z_d} \ , \qquad z_d = \frac{\tau_{s0}^2}{|\partial^2 k_s/\partial \omega^2|} \ , \qquad \nu = \frac{1}{v_p} - \frac{1}{v_s} \ .$$

Here, A_{s0} and τ_{s0} are initial amplitude and pulse length. Then

$$R_s = \frac{z_d}{z_{nl}} \ , \qquad \mu = \frac{|\partial^2 k_p/\partial \omega^2|}{|\partial^2 k_s/\partial \omega^2|} \ ,$$

$$G = \frac{z_{gr}}{z_{SRS}} \ , \qquad R_p = \frac{R_s k_p}{k_s} \ , \qquad \gamma_{p,s} = \Gamma_{p,s} z_d \ ,$$

where $z_{SRS} = 2/(G\alpha_0|A_{p0}|^2)$ is a nonlinear length scale for the SRS amplification, $z_c = \tau_{s0}/(v_p^{-1} - v_s^{-1})$ is the group delay length of pump and Stokes pulses, and $z_{nl} = n_0/(n_2 k_s \alpha_0|A_s|^2)$ is the nonlinear length scale for self-action of the radiation.

Consider soliton amplification by SRS in the field of a plane unmodulated pumping wave. Let $|q_s|^2 \gg R_s|q_{p0}|^2$. In this case the soliton evolution is described by the perturbed NLS equation

$$iq_\xi = \frac{1}{2}q_{\tau\tau} + |q|^2 q - i\gamma_s q + ig\exp(-2\gamma_p\xi)q , \qquad (2.7.3)$$

where

$$g = G|q_{p0}|^2 , \quad q = q_s .$$

This equation can be solved by considering modified conservation laws as follows. From (2.7.3), we obtain

$$\frac{d}{d\xi}\int_{-\infty}^{\infty}|q|^2 d\tau = -2\gamma_s\int_{-\infty}^{\infty}|q|^2 d\tau + 2g\int_{-\infty}^{\infty}\exp(-2\gamma_p\xi)|q|^2 d\tau . \qquad (2.7.4)$$

Substituting in this equation the ansatz

$$q(\xi,\tau) = 2\eta(\xi)\text{sech}[2\eta(\xi)\tau]\exp\{i\phi(\xi,\tau)\} \qquad (2.7.5)$$

gives an equation for the amplitude

$$\frac{d\eta}{d\xi} = -2\gamma_s\eta + 2g\exp(-2\gamma_p\xi)\eta . \qquad (2.7.6)$$

Integration gives

$$\eta(\xi) = \eta(0)\exp\left\{-2\gamma_s\xi + \frac{2g[1 - \exp(-2\gamma_p\xi)]}{2\gamma_p}\right\} , \qquad (2.7.7)$$

the length at which the soliton amplitude is easily found. Let the moment of the pump wave injection be $\eta(\xi_s) = \eta(\xi = 0)\exp(-2\gamma_s\xi_s)$. We have from (2.7.7) at $\gamma_p\xi \ll 1$:

$$\xi_R = \frac{\xi_s}{g/\gamma_s - 1} . \qquad (2.7.8)$$

This problem was also considered by *Hasegawa* (1983). He derived a formula to describe the soliton behaviour for periodically acting Raman gain. Let us consider this in more detail. Let a pump wave be injected together with soliton sequences which propagate in the fibre in both directions. The average soliton intensity I_s is

$$
\begin{aligned}
I_s &= \langle\frac{1}{2}|E_s|^2\rangle = \frac{1}{4\Delta t}\int_{-\Delta t}^{\Delta t}|E_s|^2 dt \\
&= \frac{\eta\tanh 2\eta\Delta t}{\Delta t} = \frac{\eta}{\Delta t} ,
\end{aligned} \qquad (2.7.9)
$$

where the soliton "duty cycle" $(\Delta t\, 2\eta)^{-1} \ll 1$. Then, from (2.7.1), we infer that for right- and left-propagating solitons, equations for the corresponding averaged intensities I_s^R and L_s^L read

$$\frac{dI_s^R}{dz} + 2\gamma_s I_s^R = 4\alpha I_p I_s^R \, , \tag{2.7.10}$$

$$\frac{dI_s^L}{dz} - 2\gamma_s I_s^L = -4\alpha I_p I_s^L \, . \tag{2.7.11}$$

For fused quartz, the coupling constant α for the Raman process is $\omega n_2/10c$. It also follows from (2.7.1,2) that for $z > 0$, the equation

$$\frac{dI_p}{dz} + 2\gamma_p I_p = -4\alpha I_p (I_s^R + I_s^L) \tag{2.7.12}$$

is valid. The pump wave is injected at $z = 0$. Simple rearrangement of these equations gives

$$\frac{d}{dz} \ln I_s^R + \frac{d}{dz} \ln I_s^L = 0 \, ,$$

and

$$I_p = \frac{1}{4\alpha} \frac{d}{dz} (\ln I_s^R) + 2\gamma_s \, .$$

Substituting these expressions into (2.7.12), we obtain

$$\frac{d}{dz} \left\{ \ln \frac{d}{dz} (\ln I_s^R) \right\} + 2\gamma_p = -4\alpha \left(I_s^R + \frac{I_{s0}^2}{I_s^R} \right) \, , \tag{2.7.13}$$

where

$$I_s^R = I_s^L = I_{s0} \quad \text{at } z = 0 \, , \quad \text{and} \quad I_p = I_{p0} \, .$$

Integrating gives

$$\frac{2\alpha I_{p0}}{\gamma} = \frac{1}{2\gamma} \frac{1}{I^R} \frac{dI_s^R}{dz} + \ln \left(\frac{I_s^R}{I_{s0}} \right) + \frac{2\alpha I_{s0}}{\gamma_p} \left(\frac{I_s^R}{I_{s0}} - \frac{I_{s0}}{I_s^R} \right) \, . \tag{2.7.14}$$

This equation describes the behaviour of the soliton train intensity for $z > 0$ as a function of the initial pumping intensity I_{p0} at $z = 0$, and of the linear pump wave loss $2\gamma_p$. It shows that if the pump vanishes at $z \to \infty$, the soliton intensity reaches the plateau $I_s \to I_M$. I_M is determined from

$$\frac{2\alpha I_{p0}}{\gamma_p} = \ln \left(\frac{I_{p0}}{I_{s0}} \right) + \frac{2\alpha I_{s0}}{\gamma_p} \left(\frac{I_M}{I_{s0}} - \frac{I_{s0}}{I_M} \right) \, . \tag{2.7.15}$$

For practical applications it is useful to have a simple formula that describes loss compensation due to amplification. If the distance between the amplifiers is z_0, then I_{p0} ought to be chosen so that $I_M/I_{s0} = \exp(2\gamma_s z_0)$. Then, we obtain

$$\frac{I_{p0}}{I_{s0}} = \frac{1}{2}\left(\frac{\gamma_s}{\alpha I_{s0}}\right)\left(\frac{\gamma_p}{\gamma_s}\right)(\gamma_s z_0) + 2\sinh(\gamma_s z_0) \ . \qquad (2.7.16)$$

According to an experimental estimate the parameters $\lambda = 1.5$ μm, $P_0 = 100$ mW with $\Delta t = 12$ ps ensure that solitons can be transmitted in a fibre without changing their shape, if $z_0 = 43$ km and 70 mW of cw pump power is injected in a counter-propagating direction to the soliton.

   Mollenauer et al. (1985) were the first to experimentally observe loss compensation by Raman gain. They demonstrated that at 10 ps fundamental soliton pulse with $\lambda = 1.56$ μm propagated without change of shape in a single-mode fibre of 10 km length when a counter-propagating cw Raman pump beam of wavelength λ_p was injected. The λ_p was chosen to be 1.46 μm, so that the difference of 440 cm^{-1} is close to the broad peak of the Raman gain band in quartz glass. Later, *Mollenauer* and *Smith* (1988) demonstrated the feasibility of propagating solitons without changing their shapes over distances greater than 4000 km in a fibre with periodic Raman gain. They investigated the propagation of a 55 ps soliton with $\lambda = 1.6$ μm in a fibre loop of 42 km length. The fibre was a single-mode low-loss (~ 0.22 dB/km) fibre with group-delay dispersion $D \sim 17$ ps/(nm km). Continuous wave pump power $P \sim 300$ mW was injected into the loop at a wavelength $\lambda = 1.497$ μm. They also analysed intensity pulse envelopes for 48, 96, and 125 roundtrips in the loop, corresponding to distances traversed of 2003, 4006, and 5216 km, respectively. Their analysis showed that for distances up to 2003 km, the soliton profile is remarkably unchanged. The deviation of the spectrum from that for the soliton for large distances is associated with nonlinear and stochastic processes which occur when solitons collide with a pump pulse. Raman amplification for the soliton compression is applied by *Gouveia-Neto* et al. (1987, 1988).

   Consider now a gain circuit which is based on periodically located regeneration units. The external pumping of energy is described by introducing into the right-hand side of the NLS equation a source term:

$$S = i[\exp(\Gamma\xi_0) - 1]\sum_n \delta(\xi - n\xi_0)q(\xi,\tau) \ .$$

For $\Gamma\xi_0 << 1$, one obtains

$$S = i\Gamma\xi_0\Sigma\delta(\xi - n\xi_0)q(\xi,\tau) \approx i\Gamma\int \delta(x - \xi)q(\xi,\tau)dx$$

$$= i\Gamma q(\xi,\tau) \ , \quad \xi_0 \to 0 \ . \qquad (2.7.17)$$

Thus, the dissipative term is exactly cancelled by the active term S and the soliton propagates without damping.

   The source term also leads to an emission of electromagnetic waves by optical solitons. Let us define a steady-state radiation level in the fibre, following *Malomed* (1986). To evaluate the characteristics of the radiation field, the perturbation theory for the perturbed NLS equation will be applied. First, we find the coefficient $b(\lambda)$ characterizing the continuum spectrum. As in Appendix B, the equation for $b(\lambda, t)$ has the form

$$\frac{db(\lambda, t)}{dt} = 4i\lambda^2 b + ia(\lambda) \int \left[\psi^{(2)*}(x, \lambda)^2 R(x) \right.$$

$$\left. + \psi^{(1)*}(x, \lambda)^2 R(x) \right] dx . \tag{2.7.18}$$

Here, R is a perturbation operator, $k = -2\lambda$ is the radiation wave number, and $a(\lambda)$ and $\psi(x, \lambda)$ are the Jost coefficient and Jost functions, respectively. Then, the expressions for Jost functions for single-soliton potentials, as given in Appendix B, are substituted into (2.7.18). As a result, the change in $b(\lambda)$ under the action of the n-th pump pulse is

$$\Delta_n b(\lambda, t) = -\pi\gamma\xi_0 \text{sech} \frac{\pi \left(\lambda + \frac{v}{4} \right)}{2\eta} \exp \left\{ -4i \left(\eta^2 + \frac{v^2}{16} + \frac{\lambda v}{2} \right) n\xi_0 \right\} . \tag{2.7.19}$$

In between the δ-impulses, we have

$$b(\lambda, t) = b(\lambda, 0) \exp(-\gamma\tau + 4i\lambda^2 t) . \tag{2.7.20}$$

From (2.7.19,20), one can derive a spectral density number of quanta $n(\lambda)$ (Appendix B). For this purpose, we use

$$n(\lambda) \approx \frac{1}{\pi} |b(\lambda)|^2 \quad \text{for} \quad |b|^2 \ll 1 .$$

This leads to

$$n(\lambda) \approx \pi(\gamma\xi_0)^2 \left[(\gamma\xi_0)^2 + \Psi^2 \right]^{-1} \text{sech}^2 \left[\frac{\pi}{2\eta\sqrt{2\xi_0}} \sqrt{\pi - 2\xi_0\eta^2} \right] , \tag{2.7.21}$$

where

$$2 \left[\eta^2 + \left(\lambda + \frac{v}{4} \right)^2 \right] \xi_0 = \ell\pi + \frac{\Psi}{2} ,$$

and ℓ is an integer. The spectral density inside the packet is seen to display many sharp maxima at intervals $\Delta\lambda \sim (n\xi_0)^{-1}$, the width of each maximum being

$$\delta\lambda \sim \frac{\gamma}{\eta} \ll \Delta\lambda . \tag{2.7.22}$$

The number of quanta is $N = \int_{-\infty}^{\infty} n(\lambda) d\lambda \approx 2\gamma\xi_0\eta$ and is to be less than $N_{\text{sol}} = 4\eta$.

Finally, we shall consider the influence of an extra weak wave on the soliton properties. The situation can occur, for example, in interaction between two modes in coupled fibres, or in a fibre with elliptic index profiles where the interaction of perpendicularly polarized modes exists; a weak wave is in one of the modes. The equations describing the envelopes of the pulse q_1 and the wave q_2 have the form

$$iq_{1\xi} + \frac{1}{2} q_{1\tau\tau} + |q_1|^2 q_1 + i\Gamma q_1 = q_2 \delta_1 \exp(i\kappa\xi - i\Omega\tau),$$

$$iq_{2\xi} = q_1 \delta_2 \exp(-i\kappa\xi + i\Omega\tau), \qquad (2.7.23)$$

where

$$\kappa = \left[g_2 - g_1 - \frac{\omega_2 - \omega_1}{v}\right] z_{nl}, \quad z_{nl} = \frac{\lambda_1}{\pi n_2 |A_0|^2},$$

$$\Omega = (\omega_2 - \omega_1)\left(-k'' z_{nl}\right)^{1/2}; \qquad (2.7.24)$$

$$\Gamma = \gamma z_{nl},$$

$$\delta_2 \approx \delta_1 = \frac{z_{nl}(\varepsilon_3 - \varepsilon_1)\omega_2^2}{2c^2 g_1} \frac{\langle R_1 R_2^* \rangle}{\langle |R_2|^2 \rangle} = \delta,$$

$$|\delta| \ll 1.$$

The brackets $\langle\ldots\rangle$ indicate averaging over the fibre cross-section. The expression for δ corresponds to the case of two coupled fibres, ω_2, g_2, and ω_1, g_1 are the frequency and propagation constant for the pump and carrier waves, respectively; γ^{-1} is the distance over which the electric field amplitude decreases by a factor of e due to loss in a fibre.

Let us consider the case of small coupling so that energy transfer from one mode to the other is small, and solve (2.7.23) by perturbation theory, confining our attention to first order in δ. Following *Newell* (1985), we obtain an expression for the soliton parameters:

$$q_\xi + 2\Gamma q_0 = \frac{\pi\delta|q_2|\sin\chi}{\cosh(\pi a)},$$

$$\mu_\xi = \frac{2\pi\delta a|q_2|\sin\chi}{\cosh(\pi a)}, \qquad (2.7.25)$$

$$\chi = \phi + \kappa\xi - \zeta(\Omega - \mu) - \frac{1}{2}\int_0^\xi (q_0^2 - \mu^2)\, d\xi,$$

$$q_2 = |q_2| e^{i\phi},$$

where $a = (\Omega - \mu)/2q_0$; $\mu = -\zeta_\xi$; q_0 is the soliton amplitude. Further,

$$q_s = q_0 \operatorname{sech}\left[q_0(\tau - \zeta)\right] \exp\left[\frac{i}{2}\int_0^\xi (q_0^2 - \mu^2)\, d\xi - i\mu\tau\right]. \qquad (2.7.26)$$

We obtain from (2.7.25) that a steady-state solution exists, with

$$\mu_0 = \Omega, \quad q_0 = \sqrt{\kappa + \Omega^2}. \qquad (2.7.27)$$

This solution is valid within the interval

$$0 < (\kappa + \Omega^2) < \frac{\pi^2 \delta^2 |q_2|^2}{4\Gamma^2} ,$$

and the soliton envelope then has the form

$$q_s = \sqrt{\kappa + \Omega^2} \operatorname{sech} \left[\sqrt{\kappa + \Omega^2} \, (\tau - \zeta) \right] \exp \left(\frac{i}{2} \kappa \xi - i\Omega \tau \right) .$$

For coincident frequencies $\omega_1 = \omega_2$, i.e., for the $\Omega = 0$, the resulting data agrees with Newell's results. Another procedure for compensating soliton loss was suggested by *Tajima* (1987). His idea is that the fibre parameters must be chosen so that the group velocity dispersion and the core radius both decrease along the fibre length.

In the meantime *Hasegawa* and *Kodama* (1991a, b) confirmed the stability of optical solitons in fibers under small-scale periodic amplification and periodic variation of the fiber dispersion (so-called guiding-center solitons).

2.8 Modulational Instability of Electromagnetic Waves

During the propagation of an electromagnetic wave in an optical fibre, amplitude and phase modulations of the initial wave can grow. This phenomenon is referred to as the modulational instability (MI), and its existence is associated with the mutual influence of nonlinearity and anomalous dispersion on the wave.

The modulational instability was first studied for the propagation of waves in deep water by *Benjamin* and *Feir* (1967). *Bespalov* and *Talanov* (1966) considered the MI for light beam self-focusing. *Hasegawa* and *Brinkmann* (1980), *Hasegawa* (1984), *Anderson* and *Lisak* (1986), *Schukla* and *Rasmussen* (1986) studied the MI in fibres. In particular, *Hasegawa* (1984), using numerical studies, proposed the use of the MI of a plane wave to generate periodic ultrashort pulse sequences. Below we will consider the propagation of a linearly polarized wave $A \exp(ik_0 z - i\omega_0 t)$ in an optical fibre with the higher order dispersion effects taken into account (*Anderson* and *Lisak*, 1984, 1986; *Schukla* and *Rasmussen*, 1986; *Abdullaev* et al., 1987). The governing equation is

$$iq_\xi + \frac{1}{2} q_{\tau\tau} + i\Gamma q - i\beta q_{\tau\tau\tau} + i\alpha \left(|q|^2 q \right)_\tau + \delta q_{\tau\tau\tau\tau} = 0 . \qquad (2.8.1)$$

We will first examine the nondissipative case $\Gamma = 0$. For a plane wave with time-independent real amplitude ψ_0, we have

$$q = \psi_0 \exp(i\Delta k \xi) . \qquad (2.8.2)$$

From (2.8.1), we derive

$$\Delta k = \psi_0^2 . \qquad (2.8.3)$$

To investigate the MI of the plane wave, we consider a weak perturbation of the initial state

$$q = (\psi_0 + \psi(\xi, \tau))\exp(i\Delta k\xi) , \quad |\psi| \ll \psi_0 . \tag{2.8.4}$$

Then, substituting into (2.8.1) and using equations (2.8.2,3), we obtain the following linearized equation for ϕ

$$i\psi_\xi + \frac{1}{2}\,\psi_{\tau\tau} + \psi_0^2(\psi + \psi^*) - i\beta\psi_{\tau\tau\tau} + \delta\psi_{\tau\tau\tau\tau}$$
$$+ i\alpha\psi_0^2(2\psi_\tau + \psi_\tau^*) = 0 . \tag{2.8.5}$$

Let us represent ψ as $\psi = u + iv$. Separating real and imaginary parts yields

$$-v_\xi + \frac{1}{2}\,u_{\tau\tau} + 2\psi_0^2 u + \beta v_{\tau\tau\tau} + \delta u_{\tau\tau\tau\tau} - \alpha\psi_0^2 v_\tau = 0 , \tag{2.8.6}$$

$$u_\xi + \frac{1}{2}\,v_{\tau\tau} - \beta u_{\tau\tau\tau} + \delta v_{\tau\tau\tau\tau} + 3\alpha\psi_0^2 u_\tau = 0 . \tag{2.8.7}$$

A solution can be found in the form:

$$\begin{pmatrix} u \\ v \end{pmatrix} = \begin{pmatrix} u_0 \\ v_0 \end{pmatrix} \exp[i(k\xi - \Omega\tau)] . \tag{2.8.8}$$

Here, k and Ω are the (complex) wave number and (real) modulation frequency, Substituting (2.8.8) into (2.8.6,7), we obtain a dispersion relation for k and Ω

$$k = 2\beta\Omega\psi_0^2 + \beta\Omega^3$$
$$\pm \left[\alpha^2\Omega^2\psi_0^4 + \left(\frac{\Omega^2}{2} - \delta\Omega^4\right)\left(\frac{\Omega^2}{2} - 2\psi_0^2 - \delta\Omega^4\right) \right]^{1/2} . \tag{2.8.9}$$

MI is seen to occur (in the case $\delta = 0$) when

$$\Omega^2 \le \Omega_0^2 = (1 - \alpha^2\psi_0^2)4\psi_0^2 , \tag{2.8.10}$$

i.e., the long wavelength modulations are exponentially unstable. The self-steepening effect lowers the threshold of MI relative to that of the unperturbed NLS equation case. Taking account of the term containing δ as in (2.8.9) does not essentially alter Ω_0. Note that the third-order dispersion and, similarly, higher odd orders do not affect MI (*Potasek* and *Agrawal*, 1986; *Abdullaev* et al., 1987; *Potasek*, 1987). Even-order dispersions make a non-zero contribution. The condition for MI to occur is

$$\alpha^2\psi_0^2 < 1 . \tag{2.8.11}$$

In the case $\Gamma \ne 0$, we find from the condition Im $k > \Gamma$ that MI exists at

$$\Omega^2(\Omega_0^2 - \Omega^2) > 16\Gamma^2 . \tag{2.8.12}$$

This condition can be satisfied within the range $\Omega_1^2 < \Omega^2 < \Omega_2^2$, where

$$\Omega_{1,2}^2 = \frac{\Omega_0^2}{2} \pm \sqrt{\frac{\Omega_0^4}{4} - 16\Gamma^2} \, .$$

MI of a plane wave in an optical fibre was experimentally discovered by *Tai* et al. (1986). As a source, they used a laser with $\lambda = 1.319 \ \mu m$, generating 100 ps pulses at a repetition rate of 100 MHz. The modulation period was 2 ps, so the laser radiation was quasi-continuous.

The radiation was injected into single-mode optical fibres of two types. In the first type, fibres had a dispersion coefficient $D = 2.4$ ps/(nm km) and lengths $L = 0.5$ and 1.2 km, respectively. In the second type, the fibres had $D = 3.75$ ps/nm km, $L = 0.5$ and 1 km. MI was observed in all fibres. The measurement of the output power spectrum showed that, for small P, it practically coincided with the laser radiation spectrum. For a peak power of $P = 5.5$ W two symmetric satellites are observed. Their intensity grew exponentially with the increase of the distance ξ. The self-modulation period T_m is defined as

$$T_m = \pi \sqrt{\frac{2|k''|}{g_{\max}}} \, ,$$

where $g_{\max}$ is the maximal increment. Under the experimental conditions $g_{\max} = 10^{-4} \mathrm{cm}^{-1}$ and $T_{\max} \approx 2$ ps. Thus, the MI phenomenon is useful to obtain short pulses with high repetition rates. MI may be useful for an ultrafast, all-optical fibre switch (*Islam* et al., 1987).

2.9 Generation of Periodic Pulse Sequences

The MI phenomenon treated in the previous section affords a remarkable opporturnity to generate ultrashort optical pulse chains with a high repetition rate. As already noted, a wave of a constant amplitude ($|q| = q_0$) is unstable against small perturbations with modulation wavelength $\lambda > \lambda_c = \pi/q_0$. This means that if the input signal is modulated with the period $\lambda > \pi/q_0$, then the modulation amplitude increases. Its increase leads to the self-modulation of the wave and to the formation of localized pulse sequences. Dynamics of the formation of such sequences was numerically simulated by *Hasegawa* (1984), who solved numerically the NLS equation with the periodically modulated initial condition

$$q(0, \tau) = 1 + A_M \sin(2\pi/\tau_M) \, . \tag{2.9.1}$$

Damping in the fibre was chosen to be $\Gamma = 5.18 \cdot 10^{-2}$. The simulation showed that, due to the signal evolution in the fibre, a periodic pulse sequence having a width $\tau_0 \leq 1$, was formed. The repetition pulse rate turned out to be equal to the frequency of the initial modulation and independent of the modulation depth A_m. The length within which chains are formed depends on A_m and τ_M. Let us consider a fibre with a cross-section $S = 20 \ \mu m^2$, and group dispersion $(\lambda k'')^{1/2} =$

$8.52 \cdot 10^{-17}$ sec for $\lambda = 1.5 \ \mu m^2$. A peaked structure occurs, e.g., for $A_M = 0.2$, $\tau_M = 12$ at a distance $\xi_M = 3.55$ that corresponds to the average power 106 mW, $t_M = 32.4$ psec, $z_M = 5.35$ km. This method has the disadvantage that the forming chain of subpulses always has a non-zero background intensity between the pulses, so the resulting sequence is unstable.

Dianov et al. (1989) suggested the use of adiabatic amplification of the periodically modulated cw signal to generate a chain of independent solitons in a fibre. The initial periodic signal was

$$q(0, \tau) = a \sin(\pi \tau / T) \,, \tag{2.9.2}$$

and the amplitude of the initial signal is chosen to be smaller than the MI threshold. The numerical solution of the NLS equation with linear amplification (additional term in the right-hand side of the NLS equation $\sim i\alpha q$) for initial condition (2.9.2) showed that a soliton chain of width $\tau_p = 7.05 \exp(-2\alpha L)/aT$ is generated, where L is the length over which a soliton is formed.

The above problem is reduced to the search of periodic NLS solutions. Following *Akhmediev* et al. (1985, 1989), we analyze solutions of the NLS equation with periodic initial conditions. Note that another approach was given by *Tracy et al* (1984, 1988). In the NLS equation, we will proceed with the function q_1 related to q by

$$q(\xi, \tau) = q_1(\xi, \tau) \exp(i\xi) \,. \tag{2.9.3}$$

Then, $q_1(\xi, \tau)$ preserves a smooth ξ-dependence. For q_1, we have

$$iq_{1\xi} + \frac{1}{2} \, q_{1\tau\tau} + |q_1|^2 q_1 - q_1 = 0 \,. \tag{2.9.4}$$

Its soliton solution is

$$q_1 = \frac{\sqrt{2}}{\cosh \sqrt{2}(\tau - \tau_0)} \,. \tag{2.9.5}$$

Equation (2.9.4) has for its simplest solution a complex function $q_1 \exp(i\phi)$, describing a stationary wave of a unit amplitude and an arbitrary phase. It is unstable with respect to long wavelength periodic perturbations which are increased exponentially during the wave propagation along the fibre. At the fibre input, let the wave be weakly modulated so that

$$q_1 = \left[1 + \sum_j a_j(\xi) \cos(j\kappa(\tau - \tau_0)) \right] \exp(i\phi) \,, \tag{2.9.6}$$

where $|a_j| \ll 1$, κ is a modulation frequency, j is the harmonic of the fundamental frequency present in the incident signal, τ_{0j} is the initial phase of the j-th harmonic. Substituting (2.9.6) into (2.9.4) and restricting ourselves only to terms linear in a_j, we can show that these coefficients have the form

$$a_j(\xi) = A_j \left[\frac{j\kappa}{2} + i\left(1 - \frac{j^2\kappa^2}{4}\right)^{1/2}\right] \exp(\delta_j\xi)$$

$$+ B_j \left[\frac{j\kappa}{2} - i\left(1 - \frac{j^2\kappa^2}{4}\right)^{1/2}\right] \exp(-\delta_j\xi)$$

$$= A_j \exp(i\alpha_j + \delta_j\xi) + B_j \exp(-i\alpha_j - \delta_j\xi) , \qquad (2.9.7)$$

where

$$\tan \alpha_j = \frac{2\delta_j}{j^2\kappa^2} , \qquad \delta_j = j\kappa \left(1 - \frac{j^2\kappa^2}{4}\right)^{1/2} ;$$

δ_j is the incremental increase of the perturbation of the j-th harmonic, and is real in the range $0 < j\kappa < 2$; A_j and B_j are constants and

$$|A_j \exp(\delta_j\xi)| \ll 1 , \qquad |B_j \exp(-\delta_j\xi)| \ll 1 .$$

The instability growth rate δ_1 is a maximum at $\kappa = \sqrt{2}$.

First, we investigate the solution of (2.9.4) for the case of modulation with one harmonic. To get perturbation of entirely exponential growth, we need both amplitude and phase modulation with the "depth ratio" α_1. It can be shown by direct substitution that the exact solution of (2.9.4) at $\xi \to -\infty$ has the form

$$q_1(\xi, \tau) = \left[1 - \frac{M}{N}\right] \exp(i\phi') , \qquad (2.9.8)$$

where

$$M = \frac{1}{2} p\kappa^2 \cosh \delta_1(\xi - \xi_{01}) + ip\delta_1\sinh \delta_1(\xi - \xi_{01}) ,$$

$$N = p\cosh \delta_1(\xi - \xi_{01}) - \cos \kappa(\tau - \tau_{01}) ,$$

$$p = \frac{\kappa}{\delta_1} ,$$

$$\phi' = \phi - \cos^{-1}\left(1 - \frac{\kappa^2}{2}\right) .$$

The constant ξ_{01} is defined from the relation

$$A_1 = \delta_1 \exp(-\delta_1\xi_{01}) .$$

Equation (2.9.8) shows that the initial state $\sim \exp(i\phi)$, due to the development of instability, acquires modulation whose depth increases nonlinearly up to a certain maximum for $\xi = \xi_{01}$, then it returns to the stationary solution with another phase for $\xi \to \infty$

$$q_1 = \exp[i(\phi + \Delta\phi_1(\kappa))] ,$$

i.e., the solution (2.9.8) returns the field state to the initial one with the same initial unit amplitude, but with the phase altered with respect to the initial one by an amount

$$\Delta \bar{\phi}_1(\kappa) = 2 \left[\cos^{-1} \left(\frac{\kappa}{2} - 1 \right) \right]^2 .$$

At the final stage of the process for $\xi \to \infty$, a linear term of the expansion of (2.9.8) in the small parameter $\exp[-\delta_1(\xi - \xi_{01})]$ again coincides with (2.9.6), but now $A_1 = 0$, $B_1 = \delta_1 \exp(\delta_1 \xi_{01})$ and the phase is $\phi + \Delta\phi_1(\kappa)$.

Thus, the exponentially decreasing term in (2.9.7), which is usually neglected, has a real physical meaning. For $\kappa \to 0$ the period in τ increases to ∞, and from (2.9.8), we derive the solution as a rational fraction:

$$q_1(\xi, \tau) = \left\{ 1 - \frac{4[1 + 2i(\xi - \xi_{01})]}{1 + 4(\tau - \tau_0)^2 + 4(\xi - \xi_{01})^2} \right\} \exp(i\phi') . \qquad (2.9.9)$$

The corresponding field distribution has the shape of a single "dark" pulse on a continuous signal background that changes its shape upon variation of ξ. For $\xi \to \infty$ the pulse vanishes and the initial stationary field is restored. For $\delta_1 \to 0$, a power growth occurs faster than an exponential growth of the initial perturbation. In particular, we would like to point out the case $\kappa = \sqrt{2}$, corresponding to the maximum of δ_1 when (2.9.8) turns into

$$q_1(\xi, \tau) = \frac{\cosh \sqrt{2}(\tau - \tau_{01}) + i\sqrt{2} \sinh (\xi - \xi_{01})}{\cosh \sqrt{2}(\tau - \tau_{01}) - \sqrt{2}\cosh(\xi - \xi_{01})} \exp(i\phi) . \qquad (2.9.10)$$

This transforms the initial condition $q = \exp[i(\phi'+\pi/2)]$ into $q = \exp[i(\phi' - \pi/2)]$ so that the total phase change is π.

Now, we will briefly describe results of the analysis for an initial modulation by a composite periodic signal. In the range of $0 < \kappa < 1$, not only the main perturbation mode appears to be unstable, but also its harmonic of frequency 2κ. Thus, two harmonics need to be taken into account in the initial condition. During the evolution of such a perturbation, the signal shape remains periodic in τ and depends both on the relative difference of phase of the two harmonics, and on the ratio of their initial amplitudes. We will not provide a complete solution here because it is very lengthy (Akhmediev et al., 1985). We consider only the case $\kappa = 2/\sqrt{5}$, where the incremental increase of two harmonics coincide: $\delta_1 = \delta_2 = 4/5$. The symmetric solution has the form

$$q_1(\xi, \tau) = - \left[\cosh^2 \left(\frac{4}{5} \xi \right) + P(\tau) \cosh \left(\frac{4}{5} \xi \right) - C(\tau) \right.$$

$$\left. + iF(\tau)\sinh \left(\frac{4}{5} \xi \right) \right] / \left[\cosh^2 \left(\frac{4}{5} \xi \right) \right.$$

$$\left. + \frac{5}{4} F(\tau) \cosh \left(\frac{4}{5} \xi \right) + C(\tau) \right] \exp(i\phi') , \qquad (2.9.11)$$

where

$$P(\tau) = \frac{1}{\sqrt{5}} \left[2\cos\left(\frac{2}{\sqrt{5}}\right)(\tau - \tau_{01}) - \cos\left(\frac{4}{\sqrt{5}}\right)(\tau - \tau_{02}) \right] ,$$

$$F(\tau) = \frac{4}{3\sqrt{5}} \left[2\cos\left(\frac{2}{\sqrt{5}}\right)(\tau - \tau_{01}) - \cos\left(\frac{4}{\sqrt{5}}\right)(\tau - \tau_{02}) \right] ,$$

$$C(\tau) = \frac{8}{9} + \cos\left(\frac{2}{\sqrt{5}}\right)(\tau + \tau_{01} - 2\tau_{02}) + \frac{1}{9}\cos\left(\frac{4}{\sqrt{5}}\right)(3\tau - \tau_{01} - 2\tau_{02}) .$$

The total phase change in this solution is 2π.

When the modulation frequency lies in the region $0 < \kappa < 2/n$, where n is an integer, the n-th harmonics of the main modulation frequency appear to be unstable, and n-mode separatrix solution of more complex structure can exist. Note that for $n > 2$ a slight change of initial conditions may considerably alter the subsequent evolution of the field. Let us describe the peculiarities of such solutions.

1) For n-mode separatrix solutions the first expansion term for $\xi \to \infty$ has the form of a sum of n-elementary exponentially growing perturbations of the type shown in (2.9.6), and for $\xi \to -\infty$ exponentially decreasing terms.

2) The total phase change $\Delta\phi_n(\kappa)$ of the separatrix solution for the variation of ξ from $-\infty$ to $+\infty$ is equal to the sum of the phase changes obtained from elementary solution of (2.9.8), i.e.,

$$\Delta\phi_n(\kappa) = \sum_j \Delta\phi_1(j\kappa) , \qquad (2.9.12a)$$

irrespective of the ratio of the coefficients A_j, as predetermined by the initial conditions.

3) Choosing $A_j = \delta_j$, so that all $\xi_{0j} = 0$, a complete solution will appear to be symmetrized over ξ, i.e., there is a value for the phase ϕ, so that $q_1(\xi, \tau) = q_1^*(-\xi, \tau)$. For large separated values of ξ_{0j}, the complete solution is divided into the sum of n-elementary solutions of (2.9.8). In the case when the initial state is not a separatrix trajectory, a complete solution will be multi-periodic in ξ. Below, we will report the results of numerical calculations.

Let us consider the case of simple harmonic modulation of an input signal. The pulse shape calculated from (2.9.8) for different values of κ is shown in Fig. 2.11. For $\kappa = \sqrt{2}$ in part C the inverse of the pulse duty cycle is equal to 6. The fibre length needed to obtain the narrowest pulse is about 200 m, and the average cw input power for a fibre with 20 μm^2 cross-section is 2 W. Here, in the case of simple harmonic modulation, the shape and relative pulse duration at the fibre output can be controlled by changing the mean power and modulation frequency of the continuous input signal.

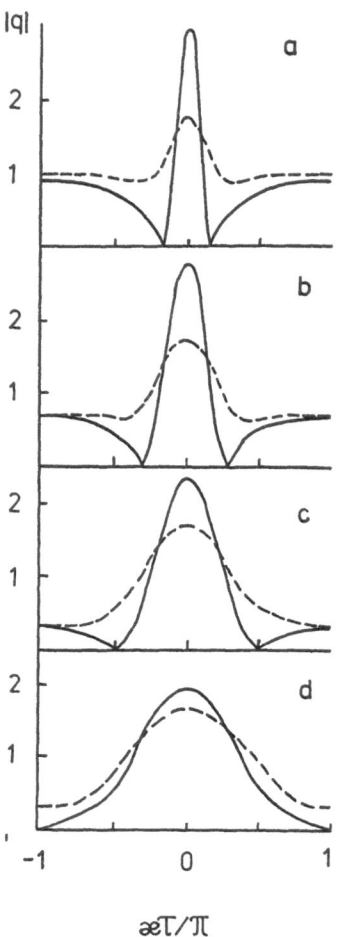

Fig. 2.11. Pulse shapes with simple harmonic modulation of the input signal. Pulse shape calculated by (2.9.8) with $\tau_{01} = 0$; solid curves: $\xi_{01} - \xi = 0$, dashed curves: $\xi_{01} - \xi = \pm 0.75$; a) $\kappa = 0$, b) $\kappa = 0.5$, c) $\kappa = \sqrt{2}$, d) $\kappa = \sqrt{3}$

The possibility to transform continuous radiation in an optical fibre into an array of short pulses opens new possibilities in fabricating fibre-based opto-electronic devices, since it is now possible to produce ultrashort pulses and to convert infrared radiation frequencies into lateral ones shifted by the modulation frequency with respect to the central frequency. Modulation can also be useful in specifying a transmission code. We note that *Its* et al. (1987) used the IST method to construct a solution which generalizes (2.9.11). It has the form

$$q(\xi) = \alpha \exp(i\varphi) \frac{\theta_{-1}(\xi, \tau)}{\theta_1(\xi, \tau)} \exp(-2i\alpha^2 \xi) , \qquad (2.9.12b)$$

where

$$\theta_1(\xi,\tau) = \sum_{k_\nu \{0,1\}^{2N}} \exp\left\{ \sum_{\nu>\mu} \ln\left[\frac{\gamma_\nu - \gamma_\mu}{\gamma_\nu + \gamma_\mu}\right] k_\mu k_\nu \right.$$

$$\left. + \sum_{\nu=1}^{2N} k_\nu \left(2i\kappa_\nu \tau - 4\kappa_\nu \lambda_\nu \xi - (l-1)\ln\frac{\gamma_\nu + 1}{\gamma_\nu - 1} + \eta_\nu \right) \right\} ;,$$

$$l = -1, 1 .\qquad\qquad\qquad (2.9.13)$$

Here θ_1 is the elliptic Jacobi theta-function,

$$\kappa_\nu = \left(\alpha^2 - \lambda_\nu^2\right)^{1/2} ,\quad \gamma_\nu = \frac{i\kappa_\nu}{\alpha + \lambda_\nu} ,\quad \nu = 1,\ldots,2N .$$

$$\eta_j = \sum_{\substack{i=l \\ j\neq l}}^{N} \ln\frac{\gamma_\nu + \gamma_\mu}{\gamma_\nu - \gamma_\mu} + c_j + id_j ,$$

$$\eta_{j+N} = \sum_{\substack{i=l \\ j\neq l}}^{N} \ln\frac{\gamma_\nu + \gamma_\mu}{\gamma_\nu - \gamma_\mu} - c_j + id_j ,$$

$$\lambda_{N+j} = -\lambda_0 ,\quad j = 1,\ldots,N .$$

$$\lambda_j : |\lambda_j| < \alpha ,\quad c_j, d_j \in R ,$$

$$\text{Im } \lambda_j = 0 ,\quad j = 1,\ldots,N .$$

Let $\delta_j = \kappa_j \lambda_j$, $\delta_1 < \ldots < \delta_N$. Then, as $\xi \to -\infty$, the solution has the asymptotic form

$$q(\xi,\tau) \simeq \alpha \exp\left[-2i\alpha^2 \xi + i\varphi_0 - 2i\sum_{j=1}^{N} \tan^{-1}\left(\frac{\kappa_j}{\lambda_j}\right) \right]$$

$$\times \left[1 + \sum_{i=1}^{N} \exp\left[4(\xi - \xi_{01})\delta_1\right] A_1 \cos\left[2\kappa_1(\tau - \tau_{01})\right] \right] ,\qquad (2.9.14)$$

where

$$t_0 = \frac{c_1}{4\delta_1} ,\quad \tau_0 = -\frac{d_1}{2\kappa_1} ,$$

$$A_1 = -\frac{4\kappa_1}{\alpha}\exp(-i\varphi_1) \prod_{\substack{j=1 \\ j\neq 1}}^{N} \frac{\gamma_1 + \gamma_j}{\gamma_1 - \gamma_j} \prod_{j=1}^{N} \frac{\gamma_{N+1} - \gamma_j}{\gamma_{N+1} + \gamma_j} ,$$

$$\varphi_1 = \tan^{-1}\left(\frac{\lambda_1}{\kappa_1}\right) .$$

When $\xi \to +\infty$, we find

$$q(\xi, \tau) \simeq \alpha \exp\left[-2i\alpha^2\xi + i\varphi + 2i \sum_{j=1}^{N} \tan^{-1}\left(\frac{\kappa_j}{\lambda_j}\right)\right]$$

$$\times \left[1 + \sum_{i=1}^{N} \exp\left[-4(\xi - \xi_{01})\delta_1\right] B_1 \cos\left[2\kappa_1(\tau - \tau_{01})\right]\right], \quad (2.9.15)$$

$$B_1 = -\frac{4\kappa_1}{\alpha} \exp(i\varphi_1) \prod_{\substack{j=1 \\ j \neq 1}}^{N} \frac{\gamma_1 + \gamma_j}{\gamma_1 - \gamma_j} \prod_{j=1}^{N} \frac{\gamma_{N+1} - \gamma_j}{\gamma_{N+1} + \gamma_j},$$

when $N = 1; 2$ and $\kappa_1 = \kappa$, these formulas reduce to those obtained above.

2.10 Multi-Mode Effects
on Soliton Regimes of Propagation

In the single-mode fibre the wave equation reduces to the one-dimensional NLS equation which describes the evolution of the complex envelope of the electric field along the fibre. However, this simple situation does not arise in multi-mode fibres because of coupling between optical fibre modes through nonlinearities and nonuniformities of different types (*Jain* and *Tsoar*, 1978). In this case the dynamics of pulses is described by a set of coupled nonlinear wave equations, derived below (*Crossiniani* et al., 1982).

Consider the propagation of an optical pulse in a multi-mode fibre. Coupling between the modes can occur due to a refractive index nonlinearity. The dielectric permeability will be assumed to be

$$\varepsilon(\varrho, z, \omega) = \varepsilon_L(\omega) + \varepsilon_2 |E_x(\varrho, z, \omega)|^2, \quad (2.10.1)$$

where ϱ is the transverse coordinate, and E_x is the electric field component perpendicular to the direction of pulse propagation. Here, $\varepsilon_L(\omega)$ is the linear part of ε, and $\varepsilon_2|E_x|^2$ can be treated as a correction to ε_L leading to mode coupling. Below, we will consider only the influence of nonlinear coupling between the modes.

The field $E(\varrho, z, t)$ will be sought as the expansion over the modes

$$E_x(\varrho, z, t) = \sum_{\nu} R_\nu(\varrho) \exp[i\omega_\nu t - ig_\nu(\omega)z]A_\nu(z, t), \quad (2.10.2)$$

where $R_\nu(\varrho)$ are the transverse modes, g_ν is the propagation constant for the ν-th mode, and $A_\nu(z, t)$ are slowly varying mode amplitudes.

Applying this expansion and using the procedure outlined in Sect. 2.1., we obtain a set of coupled NLS equations for $A_\nu(z, t)$:

$$\left(\frac{\partial}{\partial z} + \frac{\partial}{v_g \partial t} - \frac{i}{2B_\nu} \frac{\partial^2}{\partial t^2} \right) A_\nu(z,t)$$

$$= -2i \left[\sum_\mu R_{\mu\nu} |A_\mu|^2 + \frac{1}{2} R_{\nu\nu} |A_\nu|^2 \right] , \quad \nu = 1, 2, \ldots . \quad (2.10.3)$$

Here,

$$R_{\mu\nu} = \frac{\omega_0 \varepsilon_2}{2 n_L c} \frac{\iint_{-\infty}^{\infty} R_\mu^2(r) R_\nu^2(r) dx\,dy}{\iint_{-\infty}^{\infty} R_\nu^2(r) dx\,dy} ;$$

v_ν is the velocity of the ν-th mode,

$$v_\nu^{-1} = \frac{dg_\nu}{d\omega} \bigg|_{\omega = \omega_\nu} ,$$

$$B_\nu^{-1} = \frac{d^2 g_\nu}{d\omega^2} \bigg|_{\omega = \omega_\nu} .$$

Unlike the single NLS equation, the set (2.10.3) for arbitrary parameters is non-integrable. Complete integrability is feasible for a specific choice of parameters. *Manakov* (1973) showed that n-coupled NLS equations (2.10.3) are completely integrable if all velocitites v_ν and all nonlinear parameters $R_{\mu\nu}$ are equal ($v_\nu = v_0$, $R_{\mu\nu} = R_0$). Now, we analyze the set (2.10.3). Its solitary wave solutions can be obtained for the particular case when all frequencies are chosen in such a way that the different modes have equal group velocity, i.e.,

$$v^{-1} = \frac{dg_1}{d\omega} \bigg|_{\omega = \omega_1} = \frac{dg_2}{d\omega} \bigg|_{\omega = \omega_2} = \ldots . \quad (2.10.4)$$

A solitary wave solution is sought in the form

$$A_\nu(z,t) = A_{0\nu} \exp\left(\frac{iz}{2B_\nu \tau_0^2} \right) \operatorname{sech}\left(\frac{t - z/v}{\tau_0} \right) . \quad (2.10.5)$$

Substituting (2.10.5) into (2.10.3), we find that the ansatz (2.10.5) is permissible if the condition

$$-\frac{1}{B_\nu \tau_0^2} = 2 \sum_\mu R_{\mu\nu} |A_{0\mu}|^2 - R_{\nu\nu} |A_{0\nu}|^2 \quad (2.10.6)$$

is satisfied. This allows us to extend the condition for soliton existence in a single-mode fibre to the multi-mode case. In particular, assuming B_ν to be weakly dependent on ν in a multi-mode step-index fibre, we can rewrite (2.10.6) as

$$2R|A_0|^2 = -\frac{1}{B\tau_0^2} ,$$

where

$$|A_{01}|^2 = |A_{02}|^2 = \ldots = |A_0|^2 \qquad (2.10.7)$$

$$B = \left. \frac{d^2 g}{d\omega^2} \right|_{\omega=\omega_0} , \qquad R = \frac{\omega_0 \varepsilon_2}{2 n_L c} \sum_\mu |R_\mu(\varrho = 0)|^2 .$$

In deriving (2.10.7), we have used the relation

$$\frac{\sum_\mu R_\mu^2(\varrho)}{\sum_\mu R_\mu^2(0)} = \frac{n_L^2(\varrho) - n_L^2(a)}{n_L^2(0) - n_L^2(a)} , \qquad (2.10.8)$$

where a is the step-index fibre core radius.

In most cases the condition (2.10.4) is not valid. Nevertheless, one can find approximate solutions by a self-consistent method put forward by *Hasegawa* (1980). He considered the case when the velocity differences between modes are small and showed that, when the condition

$$\frac{(v_\nu - v_0)^2}{v_0^2} << -\frac{v_0^2}{B_\nu} \sum_\mu R_{\nu\mu} |A_\mu|^2$$

holds, then the mode deviating in velocity from the average value v_0 can be trapped by the main packet.

The case of two-mode fibres was numerically investigated by *Vysloukh* (1982). The initial condition was chosen to be

$$q^{(1)}(\tau, 0) = q^{(10)} \exp(i\zeta\tau) ,$$

$$q^{(2)}(\tau, 0) = q^{(20)} \exp(-i\zeta\tau) , \qquad (2.10.9)$$

$$q^{(10)} = q^{(20)} = \lambda \operatorname{sech} \lambda\tau , \quad \lambda = 2 ,$$

where q is the dimensionless amplitude, and ζ is the group delay at the dispersion length divided by the initial pulse duration. The numerical results carried out for the two cases $\zeta = 1.63$ and $\zeta = 1$ are shown in Fig. 2.12. The data show that for large values of velocity detuning, there is only partial suppression of intermode dispersion (a), while at small velocity detuning full supression is observed (b).

2.11 The Soliton Laser

The use of the soliton effects in fibres allows new perspectives in generating ultrashort pulses (USP) with tunable frequency and arbitrary duration.

Mollenauer and *Stolen* (1984) developed a device which they named the "soliton laser". Here, a fibre is used as a nonlinear feedback element in the composed laser cavity. In this manner, they obtained subpicosecond-bandwidth-limited pulses of arbitrary width.

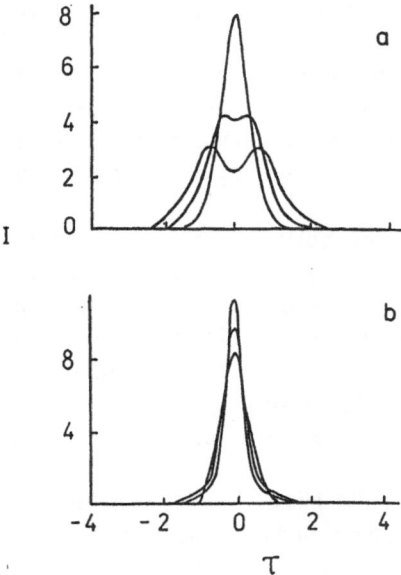

Fig. 2.12. Partial suppression of intermode dispersion during total suppression of pulse broadening (a); total suppression of intermode dispersion (b)

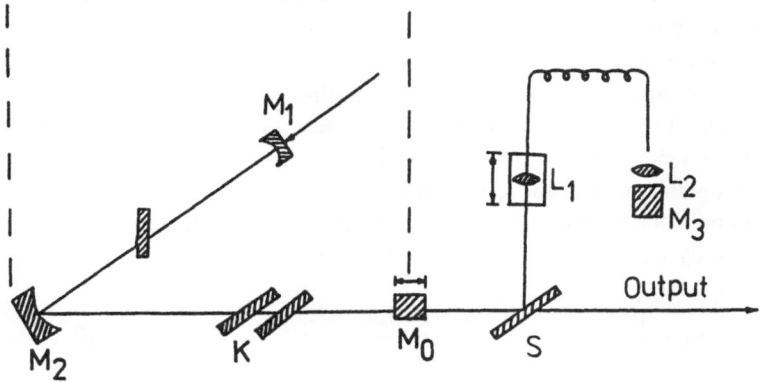

Fig. 2.13. Soliton laser. M_1, M_2, M_0 are mirrors of the first resonator; M_0, M_3, are mirrors of the second resonator; K is a birefringent plate; S is a beam splitter, L_1, L_2 are lenses

A schematic outline of the device is shown in Fig. 2.13. The soliton laser con-sists of two coupled resonators – a synchronously pumped mode-locked colour-center laser (mirrors M_1, M_2, M_0 and a birefringent plate), which is coupled to a nonlinear fibre resonator formed by the mirrors M_0 and M_3, by a beam splitter S, a microscopic lens L_1, and a fibre.

The colour-center laser alone has the following characteristics. For a pump pulse of 5 W at $\lambda = 1.064$ μm pulses with duration of ≥ 8 ps, repetition rate

100 MHz, wavelength $\lambda = 1.4 \sim 1.6$ μm, and with an average power of ~ 1 W are generated.

A polarization preserving fibre (high-birefringence) with a core diameter of 8.6 μm was used. The choice of such a fibre is motivated by the fact that, with a nonbirefringent fibre, the polarization of the pulses would fluctuate with the fibre length L, wavelength λ, and other factors, and the stable regime of self-action could then be violated.

The soliton laser works on the following principle: a laser-generated train of pulses with the wavelength at negative dispersion region travels in a single-mode fibre. One part of the laser radiation is eliminated and the other goes to a coupled resonator. In the course of travelling, the pulse, which is injected into the fibre, narrows temporally as does the multi-soliton pulse at the beginning of the compression (see Sect. 2.3.3 and Fig. 2.8). The narrowed pulse, with broadened spectrum, returns to the primary resonator of the colour-center laser and is superimposed on the pulse in that cavity.

By appropriate selection of the returning pulse phase (i.e., by selecting the length of the fibre and the additional resonator), one could obtain (after the pulses) superposition in the main resonator pulse with duration shorter than primary duration.

The process is repeated until the pulse at the fibre output acquires a stationary form. The stationary conditions mean that the pulse returning to the main resonator from the fibre has the same shape and duration as a pulse in the resonator. In other words, the pulse at the fibre output has to be the same as the pulse at the fibre input. This means that the pulse, after nonlinear evolution, is the same. This can be obtained in the case of multi-soliton pulse with integer N and when the full fibre length $2L$ is to the soliton period z_s.

In the case under consideration, the form corresponds to an $N = 2$ soliton. The working conditions for the soliton laser required for generation of the $N = 2$ soliton in a fibre by an initial pulse with a $sech^2\tau$ profile are

$$P_2 = 4P_1 , \quad P_1 = \frac{(0.776\lambda_{vac}^2/\pi^2 cn_2)|D|S_{eff}}{\tau_0^2} , \qquad (2.11.1)$$

where P_1 is the peak power, and $S_{eff} = 1.11 S_{geom}$ is the effective area of the fibre core. The second condition is that twice the fibre length should equal the soliton period z_s, i.e.,

$$2L = z_s = \frac{\pi z_d}{2} . \qquad (2.11.2)$$

The conditions for stable operation of the soliton laser can be obtained by inspecting Fig. 2.14, where the functional dependence of P_1, P_2, $\hat{P}$, and z_s on $1/\tau$ are shown (here $\hat{P}$ is the peak power in the fibre for fixed pulse energy). For stable laser operation the pulse width must satisfy the condition (2.11.2). In Fig. 2.14 this corresponds to the point τ_2. For this reason the slope of $\hat{P}$ should be chosen so that $\hat{P}$ intersects with P_2 at the τ_2. This is achieved by using the focusing lens L_1 to change the fraction of the pulse power injected into the fibre.

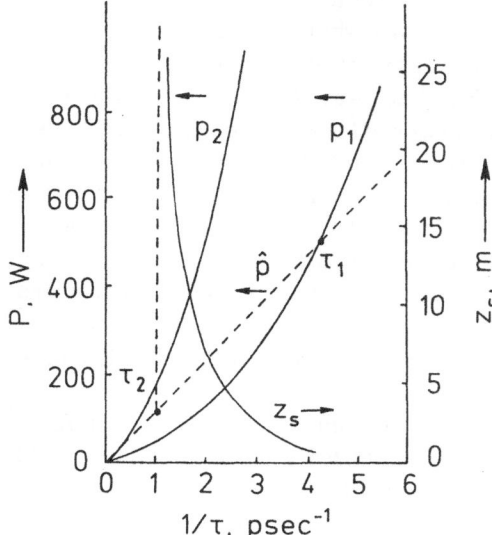

Fig. 2.14. The quantities P_1, P_2, $\hat{P}$ and z_s as functions of $1/\tau$, $\hat{P}$ is the dashed line; τ_1, τ_2 are stable operating points of the soliton laser. From *Mitschke* and *Mollenauer*, 1986

The mechanism by which the laser approaches the working point τ_2 is as follows. Let the initial pulse be too long, so that $1/\tau < 1/\tau_2$. Then, it follows from Fig. 2.14 that $\hat{P} > P_2$ and $z_s > 2L$. Both conditions result in a narrowing of the pulse, which is then injected back into the laser. After several repetitions the system moves towards the point τ_2. The measured value of P was found to be close to $P_2(\tau)$, within the bounds of experimental error.

The conditions on the fibre parameters when used in the soliton laser are not severe. Thus, the fibre length L should be $(1/2)z_s$ to within accuracy $\Delta z \sim 1$ mm, with Δz proportional to the fibre length. In contrast, the resonator length has to be determined within an accuracy of several μm.

Mollenauer and *Stolen* (1984) comment that, instead of colour-centre lasers, one could utilize semiconductor lasers, or even the fibre itself (with SRS gain). In either case the soliton laser must have a fibre resonator as a major element. The soliton laser appears to be promising for subpicosecond pulse generation.

The theory of the soliton laser has been developed by *Haus* and *Islam* (1985), *Berg* et al. (1987), and others and more recent progress has been reported by *Belanger* (1991).

2.12 The Soliton Self-Frequency Shift

A continuous shift in the mean frequency of a soliton pulse was experimentally observed by *Mitschke* and *Mollenauer* (1986). The shift is significantly larger than the spectral width of the initial pulse, and was observed in fibres after propagation distances of a few hundred meters. The effect was shown to be caused by a Raman self-pumping of the soliton which leads to the energy transfer from higher frequency components to lower frequency ones. This phenomenon was called the soliton self-frequency shift (SSFS). High-frequency pulse components act as a Raman pump source for low-frequency components. The observed frequency shift satisfied $\Delta\omega \sim P_1^2$, where P_1 is the fundamental soliton power. Taking into account that $P_1 \sim \tau^{-2}$, we have $\Delta\omega \sim \tau^{-4}$. The measured value of the frequency shift was 8 THz, which is of the order of 4% of the optical frequency ω_0. Also, the dependence of soliton self-frequency shift on the fibre length L was studied.

Gordon (1986) described the effect of the soliton self-frequency shift. In particular, he proposed the mechanism of stimulated Raman scattering as that responsible for power transfer from higher to low frequencies, and quantitatively explained the magnitude of the shift. Below, we will outline his work.

Soliton propagation in a single-mode fibre is described by the NLS equation as discussed in Sect. 2.1. In a real fibre, the soliton moves with a group velocity which depends on the mean frequency ω_0.

The Fourier transform of the soliton field is

$$\tilde{u}(\Omega) = \frac{1}{2}\operatorname{sech}\frac{\pi\Omega}{2} , \tag{2.12.1}$$

$$\Omega = (\omega - \omega_0)t_0$$

which is the soliton frequency spectrum. From (2.12.1) the soliton power $P(\tau)$ and energy spectrum $W(\Omega)$ are

$$P(\tau) = |u|^2 = \operatorname{sech}^2(\tau) ,$$

$$W(\Omega) = 2\pi|\tilde{u}|^2 = \frac{\pi}{2}\operatorname{sech}^2\frac{\pi\Omega}{2} . \tag{2.12.2}$$

Clearly, when a pulse length shortens, its spectrum broadens, and some terms which have been ignored in the NLS equation may become important. The shift of the mean soliton frequency could be caused by frequency-dependent loss or gain in the fibre. When appropriate gain and loss mechanisms are not present, the frequency shift may arise from the Raman scattering effect. This can be taken into account in the NLS equation by substituting

$$|u|^2 u \rightarrow u(t)\int ds f(s)|u(t-s)|^2 . \tag{2.12.3}$$

Here $f(s)$ is a real function, $f(-|s|) = 0$ and $\int_{-\infty}^{\infty} f(s)ds = 1$. Our main purpose is to find the soliton frequency shift by using the data for the Raman gain spectrum. We begin with the Fourier transform of a modified NLS equation

$$-i\tilde{u}_\xi(\Omega) = -\frac{1}{2}\,\Omega^2\tilde{u}(\Omega)$$

$$+ \int_{-\infty}^{\infty} d\Omega''\chi(\Omega'')\tilde{u}(\Omega - \Omega'') \int_{-\infty}^{\infty} d\Omega'\tilde{u}^*(\Omega')\tilde{u}(\Omega' + \Omega'') \;, (2.12.4)$$

$$\chi(\Omega) = \int_{-\infty}^{\infty} ds f(s)\exp(i\Omega s) \;.$$

Here, $\chi = \chi' + i\chi''$. The imaginary part χ'' is related to the Raman attenuation coefficient α_R by the expression

$$\alpha_R = 2\chi''(\Omega) \;. \tag{2.12.5}$$

Now, we may calculate the soliton self-frequency shift. The mean frequency of the soliton is equal to

$$\langle\Omega\rangle = \pi \int_{-\infty}^{\infty} d\Omega\,\Omega|\tilde{u}|^2 \;. \tag{2.12.6}$$

Using (2.12.4–6), we have

$$\langle\Omega\rangle_\xi = -\pi \int_{-\infty}^{\infty} d\Omega''\alpha_R(\Omega'') \int_{-\infty}^{\infty} d\Omega\,\Omega\tilde{u}^*(\Omega)\tilde{u}(\Omega'')$$

$$\times \int_{-\infty}^{\infty} d\Omega'\tilde{u}^*(\Omega')\tilde{u}(\Omega' + \Omega'') \;. \tag{2.12.7}$$

After the integrals have been evaluated, the mean frequency of the soliton ω_0 is governed by the equation

$$\frac{d\omega_0}{d\xi} = -\frac{\pi}{8} \int_{-\infty}^{\infty} \frac{d\Omega\,\Omega^3\alpha_R}{\sinh^2(\pi\Omega/2)} \;. \tag{2.12.8}$$

Using data for the Raman gain spectrum, *Gordon* (1986) obtained the following expression for the frequency displacement in tetraherz region $\omega = -\Omega/2\pi t_0$ (t_0 is measured in picoseconds)

$$\frac{d\omega_0}{d\xi}(\text{THz/km}) = \frac{0.0436}{\tau^4} \;. \tag{2.12.9}$$

As mentioned above, the τ^{-4} dependence was observed in experiments by *Mitschke* and *Mollenauer* (1986). This result may be obtained by using a Taylor series expansion of $\int_{-\infty}^{\infty} ds f(s)|u(\tau - s)|^2$ over τ:

$$|u|^2 - \frac{c_1\partial|u|^2}{\partial\tau} + \frac{(1/2)c_2\partial^2|u|^2}{\partial\tau^2} - \cdots \;, \tag{2.12.10}$$

where $c_n = \int_{-\infty}^{\infty} ds \, s^n f(s)$. Then, on the right side of the NLS equation, the perturbation term $ic_1(|u|^2)_\tau u$ appears. Using adiabatic perturbation theory, *Kodama* and *Hasegawa* (1987) studied the evolution of the single-soliton solution and obtained the same results as Gordon (1986). As seen from (2.12.9), the effect for picosecond pulses is neglible, but for the subpicosecond case it is large. This circumstance must be taken into account in any analysis of a soliton-based telecommunication systems. The SSFS also influences a switching phenomenon in a fibre nonlinear optical loop mirror. *Islam* et al. (1988) showed that SSFS reduces the maximum transmission to 60%, and breaks up pulses at higher powers. Femtosecond pulse-switching using the soliton self-frequency shift was proposed recently by *Blow* et al. (1989). They used the SSFS for temporal multiplexing of short pulses. This is possible because the frequency shift is inversely proportional to the pulse width in the fourth order and, thus, small changes in the pulse width lead to large changes in the frequency shift. The radiation emitted by the soliton under the action of the Raman self-frequency shift was found earlier by *Malomed* (1988).

2.13 Multi-Component (Vector) Optical Solitons

Nonlinear interaction between two polarizations arising from the tensor nature of the nonlinear susceptibility $\chi^{(3)}$ gives rise to new types of optical solitons. *Zakharov* and *Berkhoer* (1970) were the first to study this interaction for the case of perpendicularly polarized solitons in an isotropic medium. Later, *Manakov* (1973), using the IST method, studied the case when nonlinearity was induced by a certain electrostrictional mechanism. *Menyuk* (1986) explored soliton propagation in birefingent fibres for waves of two polarizations and different group velocities. When the group velocities were equal, rapidly oscillating instabilities occurred due to nonlinear polarization effects.

The pulse evolution in birefringent fibres is defined by the soliton interaction and energy exchange between two modes with different polarizations (*Doran* and *Wood*, 1988). A numerical simulation by *Blow* et al. (1987) showed that when the beat length between the modes with fast and slow polarizations is small in comparison to a soliton period, the fast mode is unstable and the power is transmitted into the slow mode. Further, work by *Christodoulides* and *Joseph* (1988, 1989) found a new class of soliton, which they referred to as vector solitons (with two field components).

We present the main results from their paper. Complex nonlinear polarization of a medium is given by the relation

$$P_{NL} = \chi^{(3)}[AE(EE^*) + BE^*(EE)] \,, \tag{2.13.1}$$

where A and B are dimensionless parameters characterizing the medium, and which satisfy they condition $A + B = 1$.

It follows from this that the equation for the dimensionless envelopes of the electromagnetic waves u, v with polarizations correspondingly parallel, to the x and y axes have the form

$$iu_\xi + \frac{1}{2}\, u_{\tau\tau} + \left(|u|^2 + A|v|^2\right) u + Bu|v|^2 e^{-4i\kappa\xi} = 0 \; , \tag{2.13.2}$$

$$iv_\xi + \frac{1}{2}\, v_{\tau\tau} + \left(|v|^2 + A|u|^2\right) v + B|u|^2 v^* e^{4i\kappa\xi} = 0 \; . \tag{2.13.3}$$

Here, κ is the normalized difference between the propagation constants for two modes. The integral for these equations is

$$\frac{\partial}{\partial\xi}\left[\int_{-\infty}^{\infty} d\tau \left(|u|^2 + |v|^2\right)\right] = 0 \; , \tag{2.13.4}$$

which implies that total energy is conserved. The modulational instability of this system was investigated by *Berkhoer* and *Zakharov* (1970). To obtain particular solutions, it is necessary to write

$$u = x(\tau)e^{-i(\mu+\kappa)\xi} \; ,$$
$$v = y(\tau)e^{-i(\mu-\kappa)\xi} \; , \tag{2.13.5}$$

which enables us to find stationary solutions for χ. Here μ is an arbitrary constant, $x(\tau)$ and $y(\tau)$ are real functions. Substitution of (2.13.5) into (2.13.3) produces

$$\ddot{x} + 2(\mu + \kappa)x + 2(x^2 + y^2)x = 0 \; ,$$
$$\ddot{y} + 2(\mu - \kappa)y + 2(x^2 + y^2)y = 0 \; . \tag{2.13.6}$$

This set of equations formally coincides with that describing the motion of a particle of a unit mass in a two-dimensional nonlinear potential. The corresponding Hamiltonian is

$$H = \frac{\dot{x}^2 + \dot{y}^2}{2} + (\mu + \kappa)x^2 + (\mu - \kappa)y^2 + \frac{(x^2 + y^2)^2}{2} \; . \tag{2.13.7}$$

One can find soliton solutions for $\kappa = 0$, taking into account that $r^2\dot{\theta}$ is then constant, when the substitution $x = r\cos\theta$, $y = r\sin\theta$ is used. Then the Hamiltonian has the form

$$H = \frac{\dot{r}^2}{2} + \frac{\ell^2}{2r^2} + \mu r^2 + \frac{r^4}{2} \; .$$

For $\ell \neq 0$ the particle-like solutions are not possible, because $H \to \infty$, when $\dot{r}$, $r \to 0$. For $\ell = 0$ a vector soliton exists:

$$x = r_0\mathrm{sech}(r_0\tau)\cos\theta_0 \; , \quad y = r_0\mathrm{sech}(r_0\tau)\sin\theta_0 \; ;$$

where $r_0^2 = -2\mu$, θ_0 is an arbitrary constant. The solution of (2.13.6) for $\kappa \neq 0$ is

$$x = \eta_1 \frac{\cosh(\eta_2 \tau) - (\eta_2/(\eta_1 + \eta_2))e^{\eta_2 \tau}}{\cosh(\eta_1 \tau)\cosh(\eta_2 \tau) - (\eta_1 \eta_2/(\eta_1 + \eta_2)^2)e^{(\eta_1 + \eta_2)\tau}} ,$$

where $\eta_1^2 = -2(\mu + \kappa)$, $\eta_2^2 = -2(\mu - \kappa)$, $\mu < 0$, $|\mu| > |\kappa|$. An analogous expression for y may be obtained by interchanging η_1 and η_2. Other solutions of (2.13.2,3), including stochastic ones are also possible. The last circumstances arise from the analysis by *Matinyan* and *Savvidy* (1982) and from *Chirikov* and *Shepelyansky* (1982), who studied spatially uniform solutions of equations for the Yang-Mills gauge fields. The corresponding values for the parameters are $E > 40\mu^2/3$, where E is the energy of the system. *Tratnik* and *Sipe* (1988) discovered the existence of the bound states of two solitons with constant and uniform orthogonbal linear polarizations. *Kath* and *Ueda* (1990), and *Muraki* (1990) have developed a Hamiltonian approach to describe the bound states of solitons in birefringent fibres. Complex dynamics of polarizations and switching phenomena were analyzed by *Wabnitz* (1988). New solitonic solutions for optical pulses propagating in birefringent fibers were found by *Acevec, Wabnitz* (1992) and *Tratnik* (1992).

3. Soliton Interaction

This chapter deals with soliton interaction in fibres and other nonlinear media. General representations of soliton interaction in integrable and nonintegrable systems, such as are found by the IST and direct methods are delineated here. Elastic interaction of two and N solitons in ideal fibres as well as inelastic interaction of solitons in real fibres are investigated. Also, quasi-periodic (finite-gap) states in optical fibres, the interaction between a soliton and a cnoidal wave, and soliton interaction in two coupled fibres are studied. We also discuss limitations on information transmission caused by soliton interaction, and the means of its suppression.

We retain in this chapter the standard notation for the NLS equation variables (except in Sects. 3.3,4,6). In describing fibre pulse dynamics, the replacement $t \rightarrow \xi$, $x \rightarrow \tau$ has been used.

3.1 General Representations
in Integrable and Nonintegrable Systems:
Direct and IST-Based Methods

To solve problems of information transmission in optical fibre systems using solitons, one needs to analyze the nature of the soliton interaction. Generally speaking, the interaction of solitons in a fibre leads to their mutual attraction and, ultimately, to their binding together. The corresponding soliton set will be detected as a single pulse and information loss will occur. The study of soliton interaction is also of importance in the description of turbulence in plasma and hydrodynamical systems (*Zakharov*, 1971), equilibrium and kinetic properties of systems in solid-state physics (*Krumhansl* and *Schrieffer*, 1975), and others.

There are several approaches now available for the description of soliton interaction. The most general one seems to be that based on direct methods (*Gorshkov* and *Ostrovsky*, 1981; *Kodama and Ablowitz*, 1984). It allows the derivation of general criteria determining the nature of soliton interaction both in integrable and nonintegrable systems. In the case of weakly interacting solitons (which will be analyzed below) the same Lagrangian can be incorporated as that used for interacting classical particles.

For the case of integrable systems, as well as for those close to being integrable, perturbation methods based on the IST for soliton theory (*Karpman* and

Maslov, 1977; *Kaup* and *Newell*, 1978) are effective. Applying these methods, we succeeded in describing inelastic processes and the wavefields radiated during soliton interaction.

Let us briefly describe the methods needed to analyze the interaction between optical solitons. To begin, we will outline a direct method, then, following *Gorshkov* and *Ostrovsky* (1981), we describe weak interaction between solitons.

The initial equation is written in the Lagrangian form

$$\frac{\partial}{\partial t}\frac{\partial L}{\partial \phi_t} + \frac{\partial}{\partial x}\frac{\partial L}{\partial \phi_x} - \frac{\partial L}{\partial \phi} = 0 \ , \tag{3.1.1}$$

where $L(\phi_t, \phi_x, \phi)$ is the Lagrangian density and ϕ is the field function.

Let us assume that (3.1.1) allows a solitary wave solution $\phi^{(0)}(\zeta = x - vt, v)$, satisfying the differential equation

$$\frac{d}{d\zeta}\left(-v\frac{\partial L}{\partial \phi_t} + \frac{\partial L}{\partial \phi_x}\right) - \frac{\partial L}{\partial \phi} = 0 \ . \tag{3.1.2}$$

In this system, weak soliton interaction occurs, provided the solitons are well separated. Moreover, their relative velocity is assumed to be small. Then, one can examine the motion of each soliton assuming it to be either in the weak field or "tail" of other solitons, and use an expansion in a small parameter to analyze soliton dynamics. According to this method, the solution is represented in the form of a series

$$\phi(x, t) = \phi^{(0)}(\zeta - s_i(t), v) + \sum_{j \neq i} \phi^{(0)}(\lambda - s_j(t), v)$$

$$+ \sum_n \varepsilon^n \phi^{(n)}(\zeta - s_i(t), \varrho, \tau) \ , \tag{3.1.3}$$

where $\varrho = \varepsilon x$, $\tau = \varepsilon t$, i is the soliton number, $ds/dt = O(\varepsilon v)$, and $\varepsilon \sim (v_i - v_j)/(v_i + v_j) \ll 1$. This perturbation method is self-consistent if we assume that the value of the field of the j-th soliton tail at the site of the i-th soliton is of the order ε^2. Substituting (3.1.3) into (3.1.1) and equating the terms of the same orders in ε^2, we obtain a set of linear equations for $\phi^{(n)}$.

For solutions to decrease for $\zeta \to \pm\infty$, it is necessary that a suitable orthogonality condition be valid. To the first order in ε, it has the form

$$\int_{-\infty}^{\infty} d\zeta \phi_\zeta^{(0)} H^{(1)} = 0 \ . \tag{3.1.4}$$

Here,

$$H^{(1)} = \frac{ds_i}{dt}\left[\frac{d}{d\zeta}\frac{\partial L}{\partial \phi_t} + \frac{d}{d\zeta}\left(-v\frac{\partial^2 L}{\partial \phi_t^2} + \frac{\partial^2 L}{\partial \phi_x \partial \phi_t}\right)\phi_\zeta^{(0)} - \frac{\partial^2 L}{\partial \phi \partial \phi_t}\phi_\zeta^{(0)}\right] \ .$$

In a similar way, we can write the condition for the absence of secular terms to the second order in ε, which results in an equation of motion for soliton centre coordinates

$$M \frac{d^2 s_i}{dt^2} = \sum_j f(s_i - s_j) . \tag{3.1.5}$$

These equations are analogous to those for classical particles with an effective mass M, and an interaction potential $U(s) \sim \int f(s) ds$:

$$M = \frac{\partial}{\partial v} \int_{-\infty}^{\infty} d\zeta \frac{\partial L}{\partial \phi_t} \phi_\zeta^{(0)} = \frac{\partial P}{\partial v} , \tag{3.1.6}$$

$$U(s_i - s_j) = \int_{-\infty}^{\infty} d\zeta \left[\frac{d}{d\zeta} \left(-v \frac{\partial L}{\partial \phi_t} + \frac{\partial L}{\partial \phi_x} \right) - \frac{\partial L}{\partial \phi} \right]_{\phi = \phi^{(0)}(\zeta - s_i)}$$

$$\times \phi^{(0)}(\zeta - s_i) , \tag{3.1.7}$$

where P is the momentum of the field.

The above-described procedure can be extended to the multi-dimensional case, where the solution depends on several variables $\zeta_1, \zeta_2, \ldots, \zeta_n$. Consider NLS equation as an example. In this case ϕ is a decreasing function of $\zeta_1 \to \infty$, and a periodic function in ζ_2. A detailed derivation of the corresponding equations for generalized P_q, U, can be found in the paper by *Gorshkov and Ostrovsky* (1981). The generalized form of (3.1.5,6) becomes

$$\left(\frac{d^2 s_i}{dt^2} \right) \frac{\partial}{\partial v_g} P_q = \sum_{j \ne i} \nabla_{s_i} U(s_{1i} - s_{1j}, \ldots, s_{mi} - s_{mj}) ,$$

$$P_q = \int_{-\infty}^{\infty} d\zeta \frac{\partial L}{\partial \phi_t} \phi_{\zeta_q}^{(0)} ,$$

$$U(s_{1i} - s_{1j}, \ldots, s_{mi} - s_{mj}) = \int_{-\infty}^{\infty} d\zeta \left[\frac{d}{d\zeta_q} \left(-v_q \frac{\partial L}{\partial \phi_t} + \frac{\partial L}{\partial \phi_{x_q}} \right) \right.$$

$$\left. - \frac{\partial L}{\partial \phi} \right]_{\phi = \phi^{(0)}(\zeta_1 - s_{1i}, \ldots, \zeta_m - s_{mi})} \phi^{(0)}(\zeta_1 - s_{1j}, \ldots, \zeta_m - s_{mj}) . \tag{3.1.8}$$

As an example, consider the interaction of solitons within the nonintegrable ϕ^4 model. The wave equation has the form

$$\phi_{tt} - \phi_{xx} - \phi + \phi^3 = 0 . \tag{3.1.9}$$

The corresponding Lagrangian is

$$L = \frac{1}{2} \left(\phi_t^2 - \phi_x^2 + \phi^2 - \frac{1}{2} \phi^4 \right) . \tag{3.1.10}$$

The solitary wave solution of this equation (the kink solution) is

$$\phi_s(x,t) = \tanh\left[\pm\frac{x-vt}{\sqrt{2(1-v^2)}}\right] . \tag{3.1.11}$$

The sign (+) corresponds to the kink, the sign (-) to the antikink solution, respectively. According to Eq. (3.1.6), the effective mass of a kink is equal to

$$M = \frac{\partial}{\partial v}\int_{-\infty}^{\infty} d\zeta\,\phi^{(0)}\phi_\zeta^{(0)} = \frac{\partial}{\partial v}\left(-\frac{v}{\sqrt{2(1-v^2)}}\right)\int_{-\infty}^{\infty}\frac{d\zeta}{\cosh^4\zeta}$$

$$= \frac{2\sqrt{2}}{3}\frac{1}{(1-v^2)^{3/2}} . \tag{3.1.12}$$

We next calculate the interaction potential for kink-kink and kink-antikink interactions, by substituting the expression for L into (3.1.7):

$$U(s_i - s_j) = \int_{-\infty}^{\infty} d\zeta\,\phi^{(0)3}(\zeta - s_i)\phi^{(0)}(\zeta - s_j) .$$

Here, it is assumed that the asymptotics $\phi^{(0)}(\zeta \pm \infty) = \phi_\pm^{(0)}$ are substracted from the kink (antikink) fields.

We thus obtain an evolution equation for the equivalent classical particle with effective mass M moving in the effective potential U

$$M\frac{d^2 s}{dt^2} = \pm 8\sqrt{2(1-v)^2}\exp\left(-\sqrt{2(1-v^2)}s\right) . \tag{3.1.13}$$

The (+) sign corresponds to the interaction of kink-kink (or antikink-antikink), (-) corresponds to the interaction of a kink-antikink pair, i.e., the kink and antikink interact by means of an exponential attractive potential. Consequently, a bound state with a kink and an antikink pair could be formed. Note that this result is valid for low values of velocity ($v < 0.2$). As shown by *Campbell* et al. (1986), at larger velocities there may be a resonant interaction between the kink and antikink, as well as some further inelastic processes (*Makhankov*, 1979).

As a second example, we examine the interaction of solitons which depend on phase variables, such as the NLS solitons. The NLS equation is obtained from the Lagrangian

$$L = \frac{i}{2}(qq_t^* - q^*q_t) + \frac{1}{2}|q_x|^2 - \frac{1}{2}|q|^4 . \tag{3.1.14}$$

Using (3.1.7), one can find equations for the components of the field pulse and the interaction potential. The NLS soliton is

$$q^{(0)} = \sqrt{2}\lambda(\text{sech}\,\lambda\zeta)\exp\left(i\frac{v}{2}\zeta + \phi\right) , \quad \zeta = x - vt ,$$

$$\phi = \omega t , \quad \lambda^2 = \omega^2 - \frac{v^2}{4} . \tag{3.1.15}$$

Substituting (3.1.15) into (3.1.8), we find the components for the field pulse and the interaction potential

$$P_\zeta = 2\pi v\lambda , \quad P_\phi = 4\pi\lambda ,$$

$$U(s_x, s_\phi) = 4\pi\lambda^3 \exp(-\lambda s_x)\cos\left[\frac{v}{2}(s_x) - s_\phi\right] . \tag{3.1.16}$$

With these components, equations of motion can be written in the form

$$\frac{d^2 z}{dt^2} = \beta\exp(z) , \quad z = u + i\psi , \quad u = -\lambda s_x , \tag{3.1.17}$$

$$\psi = -\frac{v}{2}\, s_x + s_\phi , \quad \beta = 32\lambda^2 .$$

The solutions depend on four constants and are

$$\exp\left\{\frac{1}{2}(z - z_0)\right\} = -\cosh z_\infty(t + t_R + it_I) , \tag{3.1.18}$$

where t_R, t_I, u_∞, ψ_∞ are constants. A more detailed analysis of the interaction between two NLS solitons with application to optical fibres is given in the next section.

Now, we will proceed with the consideration of methods for describing soliton interaction based on the IST (*Karpman* and *Solov'ev*, 1981; *Anderson* and *Lisak*, 1986). Consider again the problem of NLS soliton interaction. We seek a solution with well-separated solitons (*Anderson* and *Lisak*, 1986)

$$q(x, t) = q_1(x, t) + q_2(x, t) .$$

Then, soliton interaction is described by the following term

$$\varepsilon_m R_m(q_n) = i(q_m q_n^{*2} + 2q_m q_n q_n^*) , \quad m, n = 1, 2 , \quad m \neq n . \tag{3.1.19}$$

Using perturbation theory for a single NLS soliton subject to the perturbation $\varepsilon_m R_m$ (Appendix B), one can write a set of equations for the parameters of the n-th soliton

$$\frac{d\mu_n}{dt} = (-1)^n 16\eta^3 e^{-2\eta r} \cos(2\mu r + \psi) ,$$

$$\frac{d\eta_n}{dt} = (-1)^n 16\eta^3 e^{-2\eta r} \sin(2\mu r + \psi) ,$$

$$\frac{d\zeta_n}{dt} = 2\mu_n + 4\eta\, e^{-2\eta r} \sin(2\mu r + \psi) ,$$

$$\frac{d\delta_n}{dt} = 2(\eta_n^2 + \mu_n^2) + 8\mu\eta\, e^{-2\eta r} \sin(2\mu r + \psi)$$

$$+ 24\eta^2\, e^{-2\eta r} \cos(2\mu r + \psi) , \tag{3.1.20}$$

where

$$r = \zeta_1 - \zeta_2 > 0 , \quad \psi = \delta_1 - \delta_2 , \quad \eta = \frac{1}{2}(\eta_1 + \eta_2) ,$$

$$\mu = \frac{\mu_1 + \mu_2}{2} , \quad \Delta\mu \ll \mu , \quad \Delta\eta = \eta_2 - \eta_1 \ll \eta_1 ,$$

$$\eta r \gg 1 , \quad \Delta\eta r \ll 1 , \quad \Delta\mu = \mu_2 - \mu_1 .$$

Equations (3.1.20) have three constants of motion

$$\mu = \text{const} , \quad \eta = \text{const} ,$$

$$Y^2 = 16\nu^2 e^{-2\eta r} \exp\{i(2\mu r + \psi)\} = \Lambda^2 , \quad Y = \Delta\mu + i\Delta\eta . \tag{3.1.21}$$

From this, a solution is readily found

$$Y = -\Lambda \tanh(2\eta \Lambda t - \alpha_1 - i\alpha_2) . \tag{3.1.22}$$

Set $\Lambda = m + in$. Then, solution of (3.1.21) yields

$$r(t) - r(0) = \frac{1}{2\eta} \log \frac{\cosh(4\eta mt - 2\alpha_1) + \cos(4\eta nt - 2\alpha_2)}{\cosh 2\alpha_1 + \cos 2\alpha_2} , \tag{3.1.23}$$

$$\psi(t) - \psi(0) = -2 \tan^{-1} [\tanh(2\eta mt - \alpha_1) \tanh(2\eta nt - \alpha_2)]$$
$$+ 2 \tan^{-1} [\tanh \alpha_2] - 2\mu(r - r(0)) . \tag{3.1.24}$$

An analysis of these equations will be given in the next section.

The IST is also effective in the description of inelastic multi-particle effects accompanied by linear wave radiation.

The main difference between the soliton interaction in the nearly integrable perturbed NLS system as compared with, for example, the SG system, is that the binding energy of the two-soliton state of the former is zero. The two-soliton state is unstable with respect to small perturbations. However, there might be situations when perturbation stabilizes a two-soliton state (Sect. 3.3). This enables us to consider further multi-particle effects for optical solitons in fibres, such as the interaction between a two-soliton state and a soliton.

3.2 Interaction Between Optical Solitons

One of the characteristic features of solitons is their stability agains a fairly broad class of perturbations and their elastic interactions upon colliding. A consequence of applying the IST to the study of interaction of envelope solitons in an ideal single-mode fibre is that the solitons collide elastically (*Zakharov* et al., 1980). The soliton parameters do not undergo changes, but simply acquire a phase shift Δ_m (faster soliton) or $-\Delta_p$ (slower soliton)

$$\Delta_{m,p} = \frac{1}{2\eta_{m,p}} \ln \left| \frac{\lambda_m - \lambda_p^*}{\lambda_m - \lambda_p} \right|^2 , \quad \text{Re}\{\lambda_n\} < \text{Re}\{\lambda_p\} .$$

For nearly equal velocities the solitons form a bound state.

Gordon (1983) derived expressions describing the soliton interaction using the exact formulae of the IST for the two-soliton solution. Assuming that the distance between solitons was large, he found that the interaction forces exponentially decrease with the distance between the solitons, and depend only on the relative phase. This is constant with the analysis based on the direct perturbation theory for solitons (*Gorshkov* and *Ostrovsky*, 1981), and with that using the IST (*Karpman* and *Solov'ev*, 1981) (see also previous section). Gordon showed that with an initial state in the form

$$q_0 = \text{sech}(x - \xi_0) + \text{sech}(x + \xi_0) , \quad \xi_0 >> 1 \tag{3.2.1}$$

the distance between solitons r is equal to

$$r = r_0 + 2 \ln | \cos[2t \exp(-\xi_0)]| . \tag{3.2.2}$$

It follows that the distance r has an oscillating character and that a bound state of the solitons exists. Numerical simulations (*Blow* and *Doran*, 1983; *Hermansson* and *Yervick*, 1983) showed that a nonlinear interaction between solitons leads to an attraction between pulses for any value of ζ_0, with relative phase equal to zero.

We would now like to treat the interaction of two solitons using the system of equations for soliton parameters (3.1.20) obtained in the previous section. A general solution is given by (3.1.23). In particular, for solitons initially at rest and with the same amplitude, this solution simplifies to

$$r(t) = r(0) + \frac{1}{2\eta} \ln \left[\frac{1}{2}(\cosh 4\eta m t + \cos 4\eta n t) \right] , \tag{3.2.3}$$

where

$$m = -4\eta \, e^{-\eta r(0)} \sin \frac{\psi(0)}{2} ;$$
$$n = 4\eta \, e^{-\eta r(0)} \cos \frac{\psi(0)}{2} . \tag{3.2.4}$$

We will analyze this solution for two cases:

1) The initial soliton phases are equal to zero [$\psi(0) = 0$]

Then, $m = 0$, $n = 4\eta \exp[-\eta r(0)]$. The relative distance between the soliton is

$$r(t) = r(0) + \frac{1}{\eta} \ln | \cos(2\eta n t)| , \tag{3.2.5}$$

and thus the solitons undergo oscillatory motion. For $2\eta = 1$ this expression coincides with (3.2.2). The coordinate $r(t)$ is zero at time

$$t_c = \frac{1}{2\eta|n|} \cos^{-1}[\exp -\eta r(0)] . \tag{3.2.6}$$

Since $\eta r(0) >> 1$, we have

$$t_c = \frac{\pi}{16\eta^2} e^{\eta r(0)} . \tag{3.2.7}$$

2) The initial phase difference is $\psi(0) = \pi$

Then $n = 0$, $m = -4\eta \exp(-\eta r(0))$, and

$$r(t) - r(0) = \frac{1}{\eta} \ln[\cosh(2\eta mt)] . \tag{3.2.8}$$

For $\psi(0) = \pi$ the distance between the solitons is seen to increase monotonically. The separation between them is doubled after a time

$$t_d = \frac{1}{2\eta m} \sinh^{-1}[\exp(\eta r(0))] . \tag{3.2.9}$$

For $\eta r(0) >> 1$, we have

$$t_d = \frac{\eta r(0)}{8\eta^2} e^{\eta r(0)} . \tag{3.2.10}$$

As seen in Fig. 3.1, there are three different situations, namely: a monotonic increase of the soliton separation, a quasi-periodic oscillation which tends to the mode of monotonic increase, and an oscillatory mode. Hence, by appropriate choice of initial phase difference, one can suppress binding of solitons.

The interaction between optical solitons has been experimentally observed (*Mitschke* and *Mollenauer*, 1987). Both repulsion and attraction regimes appear to have been seen. They depend on the value of the soliton phase difference. In the experiment, solitons were generated using a soliton laser. The pulses first passed through a Michelson interferometer, giving a pair of pulses. The change of length of one arm of the interferometer permitted a change in the distance between the pulses from zero up to several picoseconds. The soliton phase difference was also measured.

In the experiment a polarization preserving 340 m low-loss (> 0.3 dB/km) fibre was used, with $D = 14.5$ ps/(nm km) at an operating wavelength of 1.52 μm. The pulse duration was $\sim$ 1 ps, so that the fibre length is equal to 10 soliton periods. Such a long fibre enables them to clearly observe interaction between two well-separated pulses (of the order of three pulse widths).

The results of the measurements are presented in Fig 3.2. for both repulsive and attraction regimes. As can be seen, the observed behaviour is different from that predicted by theory for the part of the curve describing attraction, and in the region where there is a small initial distance between the pulses. According to theory, two interacting solitons pass through each other. This is presented in

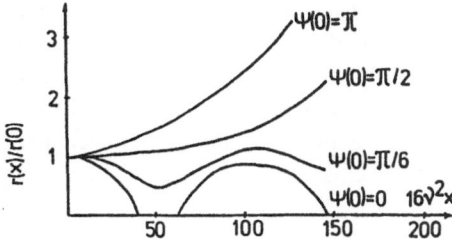

Fig. 3.1. The dependence of distance r between solitons on the initial phase difference

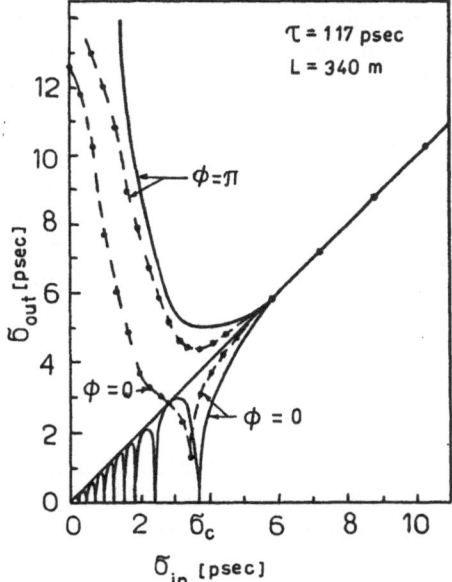

Fig. 3.2. Pulse separation at output (σ_{out}) vs. the pulse separation at input (σ_{in}) in a 340 m fibre. Solid curve: theory; dashed line: experiment. Without interaction, all points would fall on the 45° line. From *Mitschke* and *Mollenauer*, 1987

Fig. 3.2 by the oscillating structure of the theoretical curve. The experiment does not reveal such a behaviour. In the unstable region the attractive force becomes repulsive as a result of the influence of the Raman self-frequency shift.

Mollenauer and *Smith* (1989) discovered that between optical solitons at long distances (more than 1000 km) a long-range phase-independent interaction exists. *Dianov* et al. (1990) showed that the electrostrictional mechanism may be responsible for the observed anomalous interaction. *Reynaud* and *Barthelemy* (1990) observed the interaction of spatial optical solitons between two soliton beams. These authors report the existence of a mutual interaction between two close propagating soliton beams which depends on the relative phase and the distance between the beams.

In conclusion, we note that *Vysloukh* and *Cherednik* (1986) have recently developed an effective method based on a combination of the IST and numerical methods to describe soliton interaction, and to follow the evolution of an arbitrary pulse. The method is useful for studying the dynamics of multi-soliton states, and for studying the interaction of optical solitons in nearly integrable systems.

3.3 Inelastic Soliton Interactions

Close to the zero in the second-order dispersion, one has to take into account the contribution of terms arising from higher-order dispersion, as well as the action of nonlinear dissipation terms. The influence of these effects on the dynamics of a single soliton was analyzed in Chap. 2. These factors also have a nontrivial effect on the two-soliton state, and on processes such as soliton interactions, where collisions become inelastic.

Here, we will study the influence of these effects on interactions of optical solitons. The effect of the third-order dispersion (*Chu* and *Desem*, 1984), and of periodic amplification on the dynamics of the two-soliton complex was investigated numerically by *Desem* and *Chu* (1987). A distortion parameter B, defined by

$$B = \frac{\int_{-\infty}^{\infty} |q_N - q(t,x)| dx}{\int_{-\infty}^{\infty} |q(t,x)|}$$

was obtained. Here, $q(t,x)$ is the analytic two-soliton solution (2.3.19), and q_N represents solitons after a single period of amplification. The calculation showed that there are two regions of instability in the dependence of the B-parameter on propagation length in the optical fibre. The instability regions are associated with resonances which occur either between the interaction period of the solitons or in a period of the soliton phase, whithin the amplification period. To reduce this effect, it is necessary to inject solitons with unequal amplitude into the fibre. Joint action of higher-order dispersion, self-steepening and Raman scattering were studied numerically by *Hodel* and *Weber* (1987). It was shown that in the femtosecond region the Raman term is the most important and leads to a decay of the higher-order solitons.

A theoretical investigation of this problem is more difficult than that for a single soliton case. Here, we review results obtained by *Malomed* (1985) and by *Kodama* and *Nozaki* (1987). Before entering the discussion attention should be drawn to the fact that *Kodama* and *Hasegawa* (1991) have recently studied in detail the effect of an initial overlap between two solitons of different carrier wavelengths.

Kodama and *Nozaki* (1987) examined the mechanism by which a two-soliton state splits into individual solitons under the perturbing action of the Raman term and higher-order dispersion. They took a two-soliton initial pulse

$$q(x,0) = \mathrm{sech}\left(x - \frac{x_{\mathrm{in}}}{2}\right) + \mathrm{sech}\left(x + \frac{x_{\mathrm{in}}}{2}\right)\exp(i\phi) , \tag{3.3.1}$$

where x_{in} is the pulse separation, and ϕ is a phase difference. The influence of the Raman term $\tilde{\gamma}q(|q|^2)_x$ on the two-soliton state can be described by perturbation theory based on the IST (Sect. 3.2). When $\tilde{\gamma} = 0$, the IST theory predicts that solitons overlap and oscillate near $x = 0$ with period $t_0 = 4\pi/(\eta_1^2 - \eta_2^2)$. Application of perturbation theory to the soliton state leads to the following equation for the amplitude and velocity of each soliton (at $\phi = 0$):

$$\frac{d\lambda_L}{dt} = -\tilde{\gamma}\int_{-\infty}^{\infty}(q\phi_{2L}^2 + q^*\phi_{1L}^2)\frac{\partial|q|^2}{\partial x}\,dx , \quad L = 1,2 , \tag{3.3.2}$$

where $\lambda_L = (\mu_L + i\eta_L)/2$, and ϕ_{jL} are the Jost functions corresponding to the two-soliton state. We note that the soliton amplitudes are not charged by the soliton collision. The integrals in (3.3.2) were calculated numerically. The calculation showed that the velocities μ_1 and μ_2 changed in a complementary manner.

Consider also the behaviour of a two-soliton state of the NLS equation subjected to a nonlinear damping (*Malomed*, 1985):

$$iq_t + q_{xx} + 2|q|^2 q = -i\varepsilon R(q) = -i\varepsilon|q|^{2N}q . \tag{3.3.3}$$

First, we describe a two-soliton solution of the unperturbed NLS equation. Let a two-soliton state consist of solitons with zero velocities and unequal amplitudes η_1 and η_2. Henceforth, the difference

$$\kappa = \frac{\eta_1 - \eta_2}{\eta_1 + \eta_2}$$

is assumed to be small:

$$0 < \kappa\eta << \eta = \frac{1}{2}(\eta_1 + \eta_2) . \tag{3.3.4}$$

Then, the two-soliton solution has the form

$$\begin{aligned}
q(x,t) = &-4\eta\kappa|\gamma_1\gamma_2|^{1/2}(\gamma_1^* + \gamma_2^*)\big[|\gamma_1 - \gamma_2|^2 \\
&+ 2\kappa^2|\gamma_1\gamma_2|\cosh^{-1}(4\eta x)\big]\Big\{\big[1 + \gamma_1^*\gamma_2^*(\gamma_1 + \gamma_2) \\
&\times |\gamma_1\gamma_2|^{-1}(\gamma_1^* + \gamma_2^*)^{-1}\big]\cosh(2\eta x) \\
&+ \big[1 - \gamma_1^*\gamma_2^*(\gamma_1 + \gamma_2)|\gamma_1\gamma_2|^{-1}(\gamma_1^* + \gamma_2^*)^{-1}\big]\sinh(2\eta x)\Big\} , \tag{3.3.5}
\end{aligned}$$

where $\gamma_1(t)$, $\gamma_2(t)$, the complex soliton amplitudes, are time-dependent

$$\gamma_n(t) = \gamma_n(0)\exp(-4i\eta_n^2 t) , \quad n = 1,2 . \tag{3.3.6}$$

The values $(2\eta_n)^{-1}\ln[\gamma_n(0)]$ correspond to the soliton centres.

We now describe the internal structure of the two-soliton state. A maximum of $|q(x,t)|$ occurs at the point

$$x_1 = -\frac{1}{4\eta} \ln \left[\left| \frac{\gamma_1 + \gamma_2}{\gamma_1 \gamma_2} \right|^2 \kappa^{-2} \right] \, , \tag{3.3.7}$$

whereas a minimum occurs at

$$x_2 = -\frac{1}{4\eta} \ln \left[|(\gamma_1 + \gamma_2)|^2 \kappa^{-2} \right] \, . \tag{3.3.8}$$

At the maximum $|q| \sim 1$, and at the minimum $|q| \sim \kappa^2$.

Equation (3.3.8) shows that the state size is large, i.e.,

$$L = |x_1 - x_2| = \frac{1}{4\eta} \ln \left[\left| \frac{(\gamma_1 + \gamma_2)^2}{\gamma_1 \gamma_2} \right|^2 \kappa^{-4} \right] \, , \tag{3.3.9}$$

so that we are considering the case of two weakly overlapping solitons. The period of internal oscillations T is equal to the period of oscillation of the function $|\gamma_1(t) + \gamma_2(t)|^2$, or

$$T = \frac{\pi}{8\eta^2 \kappa} \, . \tag{3.3.10}$$

As seen from (3.3.9), the oscillation amplitude ΔL of the two-soliton state size is

$$\Delta L = \frac{1}{2\eta} \ln \left[\frac{|\gamma_1| + |\gamma_2|}{|\gamma_1| - |\gamma_2|} \right]^2 \, . \tag{3.3.11}$$

The expressions (3.3.6–9) and (3.3.11) are inapplicable when $|\gamma_1(t) + \gamma_2(t)|$ is small, i.e., for $|\gamma_1(t) + \gamma_2(t)| \leq \kappa$. This occurs when γ_1 and γ_2 are of the same order. Then, the two-soliton state has a more complex form than shown in (3.3.5):

$$
\begin{aligned}
q(x,t) = &- 4\eta \exp(-2\eta x) \Big\{ \kappa^{-1} \left[(\gamma_1^* + \gamma_2^*) - (\gamma_1^*)^2 (\gamma_1 + \gamma_2) \right. \\
&\times \exp(-4\eta x) \left] + |\gamma_1|^2 (1 + 2\eta x) \exp(-4\eta x) \right] \Big\} \\
&\times \left[1 + \kappa^{-2} |\gamma_1 + \gamma_2|^2 \exp(-4\eta x) + 2|\gamma_1|^2 (1 + 8\eta^2 x^2) \right. \\
&\times \exp(-4\eta x) + 2|\gamma_1|^2 (1 + 8\eta^2 x^2) \exp(-4\eta x) \\
&+ |\gamma_1|^4 \exp(-8\eta x) \Big]^{-1} \, .
\end{aligned} \tag{3.3.12}
$$

Here and in the following, we imply that the coordinate x is displaced by the value $(4\eta)^{-1} \ln |\gamma_1 \gamma_2|$. Expression (3.3.12) holds in the region $\eta|x|\kappa \ll 1$. The region of interest is $\eta|x| \leq \ln \kappa^{-1}$. Equation (3.3.12) describes the situation when solitons are slightly overlapped.

Once the properties of the unperturbed NLS two-soliton state have been described, we will proceed with the investigation of its evolution under perturbation

$$R = |q|^{2N} q \, . \tag{3.3.12a}$$

The variation of soliton parameters μ_n, η_n ($v_s = -4\mu$) is

$$\frac{d}{dt}(\mu_n + \eta_n) = 2i\eta_n \frac{\gamma_n}{\kappa} M_n ,\qquad (3.3.13)$$

where M_n is a perturbation matrix element,

$$M_n = \int_{-\infty}^{\infty} \left\{ \left(\Psi_n^{(1)}(x) \right)^2 \varepsilon R(q(x)) - \left(\Psi_n^{(2)}(x) \right)^2 \varepsilon^* R^*(q) \right\} dx .\qquad (3.3.14)$$

Here, $\Psi_n^{(1,2)}$ are Jost components evaluated at the spectral parameter $\lambda = \lambda_n = \mu_n + i\eta_n$:

$$\Psi_n^{(1)} = \Delta_n^{-1} D_n^{(1)}(x) e^{-\eta x} ,$$
$$\Psi_n^{(2)} = i\Delta_n^{-1} D_n^{(2)}(x) e^{-\eta x} ,\qquad (3.3.15)$$

where

$$\Delta = 1 + \kappa^{-2}|\gamma_1 + \gamma_2|^2 e^{-4\eta x} + |\gamma_1 \gamma_2| e^{-8\eta x} ,\qquad (3.3.16)$$

$$D_1^{(1)} = 1 + \kappa^{-1} e^{-4\eta x} \gamma_2(\gamma_1^* + \gamma_2^*) ,$$
$$D_2^{(2)} = \kappa^{-1}(\gamma_1^* + \gamma_2^*) e^{-4\eta x} + \gamma_1 |\gamma_1|^2 e^{-6\eta x} .\qquad (3.3.17)$$

Values for $D_2^{(1,2)}$ follow from $D_1^{(1,2)}$ by the replacement $\gamma_1 \Rightarrow -\gamma_2$, $\gamma_2 \Rightarrow -\gamma_1$.

Apart from the parameters η_n and μ_n, the magnitudes $\gamma_n(t)$ also vary gradually under perturbation. The corresponding equations are

$$\frac{d\gamma(t)}{dt} = \frac{2\gamma_n^2(0)}{\kappa} \int_{-\infty}^{\infty} \left\{ \varepsilon \frac{\partial}{\partial \lambda} \left[\Psi^{(1)}(x, \lambda) \right]^2 R(q(x)) \right.$$

$$\left. - \varepsilon^* \frac{\partial}{\partial \lambda} \left[\Psi^{(2)}(x, \lambda) \right]^2 R^*(q(x)) \right\} \Bigg|_{\lambda = \lambda_n} dx .\qquad (3.3.18)$$

We will consider hereafter the case when $|\gamma_1 + \gamma_2| \leq \kappa$. The equations (3.3.16,17) then are simplified to

$$\Delta = 1 + \kappa^{-2}|\gamma_1 + \gamma_2|^2 \exp(-4\eta x) + 2|\gamma_1|^2(1 + 8\eta^2 x^2)$$
$$\times \exp(-4\eta x) + |\gamma_1|^4 \exp(-8\eta x) ,\qquad (3.3.19)$$

$$D_1^{(2)} = D_1^{(1)} = 1 - \kappa^{-1} \exp(-4\eta x)\gamma_1(\gamma_1^* + \gamma_2^*) + |\gamma_1|^2(1 + 4\eta x) \exp(-4\eta x) ,$$
$$D_1^{(2)} = D_2^{(2)} = \kappa^{-1}(\gamma_1^* + \gamma_2^*) \exp(-2\eta x) + \gamma_1^*(1 - 4\eta x) \exp(-2\eta x)$$
$$+ \gamma_1^* |\gamma_1^2| \exp(-6\eta x) .\qquad (3.3.20)$$

As a result, we obtain an equation for the distance between the soliton centres $\Delta\xi = \zeta_1 - \zeta_2$, and another one for the relative amplitude $2\kappa\eta = \eta_1 - \eta_2$. Substituting Eq. (3.3.5) into (3.3.12a), the latter and (3.3.15–17) into (3.3.13,14), and then into (3.3.13) and (3.3.14) gives

$$\frac{d}{dt}(\Delta\zeta + 2i\eta\kappa) = (4\eta)^{2N+1} \frac{N!(N+1)!}{(2N+2)!} |\gamma_1 + \gamma_2|^{-2}$$

$$\times \left[2\mathrm{Im} \left\{ (\gamma_1 \gamma_2^*) + i\left(|\gamma_1^2| + |\gamma_2^2|\right) \right\} \right] \mathrm{Re}\{\varepsilon\} ;.\qquad (3.3.21)$$

An entirely imaginary value for ε corresponds to a conservative perturbation, whereas a real value corresponds to a dissipative one. Equation (3.3.21) indicates that an imaginary ε does not affect the two-soliton parameters. This is as expected, since terms with $N = 0$ and $N = 1$ do not violate the NLS integrability.

The dissipative case, (3.3.19) is applicable for $t \leq (\eta^{2N}|\varepsilon|)^{-1}$. This time is assumed to be much longer than the period of internal oscillations, i.e.,

$$|\varepsilon| << \eta^{-2(N-1)\kappa} .$$

Then, the equations can be averaged over the unperturbed fast oscillations, so that

$$\frac{d\Delta\zeta}{dt} = 0 ,$$

$$\frac{d\kappa}{dt} = -2(4\eta)^{2N} N!(N + 1)! \left(|\gamma_1^2| - |\gamma_2^2|\right) \mathrm{Re}\{\varepsilon\}/(2N + 2)!|\gamma_1 + \gamma_2|^2 .$$

This implies that when $|\gamma_1| = |\gamma_2|$, a two-soliton state remains stable in the sense that the difference of parameters of the two solitons which form the set remain unchanged. In particular, the most stable state occurs at a maximal overlap of the solitons.

Consider the case $N = 0$. Then,

$$\frac{d\kappa}{dt} = \mathrm{const}\, \kappa^{-1}\mathrm{Re}\{\varepsilon\} .$$

In the exactly integrable case the binding energy is zero for a two-soliton set. However, in the presence of a dissipative perturbation, the binding energy is no longer zero. To illustrate this, we consider the collision between two solitons with equal amplitudes η and slow velocities $\pm v$. We have

$$\frac{d(v^2)}{dt} = (4\eta^2)^{2N+1} \frac{N!(N + 1)!}{(2N + 2)!} [\cosh(4\eta\tau) + \cosh \Psi]^{-1} \sin \Psi \, \mathrm{Re}\{\varepsilon\} ,$$

where $\tau(t) = \int_0^t v(t')dt'$, and Ψ is the relative phase of the internal oscillations $(-\pi \leq \Psi \leq \pi)$ in the system.

It follows from this that the incremental velocity change after the collision is

$$\Delta v = \mathrm{Re}\{\varepsilon\}(4\eta)^{2N} \frac{N!(N + 1)!\psi}{(2N + 2)!} v .$$

This result is nontrivial because $\Delta v \sim \eta$. In the single soliton case, dissipation changes the amplitude rather than the velocity.

A threshold velocity v_{th} is defined by the condition $v = -v_{\mathrm{th}}$. Thus, we have

$$v_{\mathrm{th}} \sim \eta^N \left(|\mathrm{Re}\{\varepsilon\}|\right)^{1/2} .$$

This estimate is valid if the difference between the amplitudes of the colliding solitons is small

$$\Delta\eta \leq \eta^N \left(|\text{Re}\{\varepsilon\}|\right)^{1/2} .$$

The collision process between solitons in a birefringent fibre was considered using similar techniques by *Malomed* (1991) and *Cagliotti* et al. (1989).

3.4 Interaction Between a Soliton and a Nonlinear Periodic Wave in an Optical Fibre

In this section, we consider the problem of the interaction between a single soliton pulse and a periodic array of pulses in a fibre; this is the simplest case. The results reported here were obtained by A. Its and R. Bikbaev (*R.Bikbaev*, Ph.D thesis, Leningrad, 1987). As is well-known, the NLS equation has a solution of soliton and of cnoidal wave type. The latter is the simplest (finite-gap) periodic solution. We describe the interaction of these two solutions.

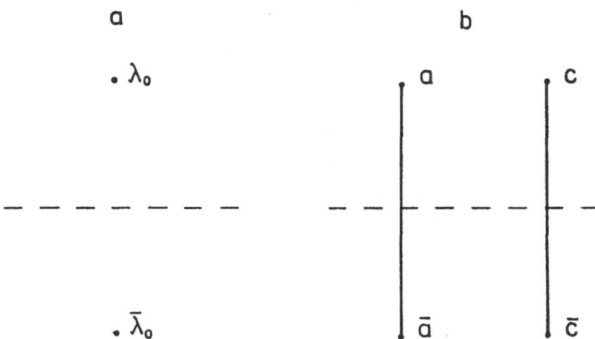

Fig. 3.3. Spectrum of a soliton (a) and cnoidal wave (b) in a complex λ-plane; a, $\bar{a}$, c, $\bar{c}$ are the branching points

Cnoidal waves belong to the class of so-called finite-gap solutions (Appendix D). N-soliton and finite-gap solutions are obtained when two types of spectral problems for the Zakharov-Shabat operator are considered. In these two cases the solutions are uniquely (up to some arbitrary constants) determined by their spectra. The spectral patterns of soliton and cnoidal waves are shown in Fig. 3.3. It is natural to assume (as is usually done with the IST) that the overlap of these two spectral patterns corresponds to the NLS solution describing the soliton-cnoidal wave interaction (*Bikbaev* et al., 1984). We obtain this solution as a degenerate case of the finite-gap solution with the following spectral data (Fig. 3.4).

By the degenerate case, we mean that we try to find a solution which is a limit of a finite-gap solution when λ_1 tends to λ_0. Thus, in Fig. 3.5 the points λ_0, λ_1 coincide.

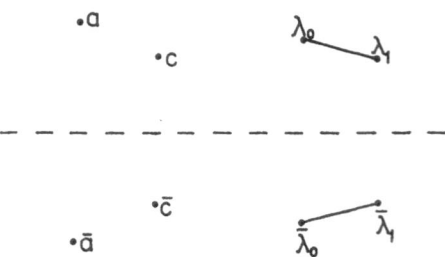

Fig. 3.4. Spectrum of the solution in the complex λ-plane describing the interaction of a cnoidal wave with a soliton

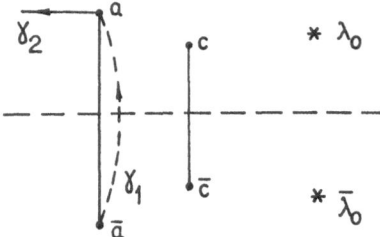

Fig. 3.5. Spectrum of the solution in the λ-complex plane to be sought

It is useful to note that the simplest case of the method is that for the spectral pattern of a cnoidal wave, and letting c tend to a. The corresponding limit solution is a soliton. This is easily seen by using the formulae of Sect. 2.3.

The NLS solution corresponding to Fig. 3.4 is (Appendix D)

$$\hat{q}(x,t) = \frac{\theta(g(x,t) - r)}{\theta(g(x,t) + r)} \exp(iE_0 x + iN_0 t + A_0) \, ,$$

where θ is a three-dimensional theta-function. Let us set $\lambda_1 \to \lambda_0$ and follow the decomposition of this θ-function. We omit the details here (see Appendix D) and present only the final result. A solution corresponding to the spectrum in Fig. 3.5 has the form

$$q(x,t) = \frac{\theta_-(x,t)}{\theta_+(x,t)} \exp[i(E_0 x + N_0 t + A_)] \, . \tag{3.4.1}$$

Quantities appearing in (3.4.1) are defined as follows:

$$\begin{aligned}
\theta_\pm &= \theta_3(\alpha_\pm) + \theta_3(\alpha_\pm - B_{21}) \exp\{-i(Vx - Wt) + \eta_\pm\} \\
&+ \theta_3(\alpha_\pm - \bar{B}_{21}) \exp\{i(\bar{V}x - \bar{W}t) + \bar{\eta}_\pm\} \\
&+ \theta_3(\alpha_\pm - (B_{21} + \bar{B}_{21})) \exp\{-i[(V - \bar{V})x \\
&- (W - \bar{W})t] + (\eta_\pm + \bar{\eta}_\pm) - 2\pi i B_{32}\} \, .
\end{aligned} \tag{3.4.2}$$

In these formulae $\theta_3(\alpha_\pm)$ are elliptic Jacobi theta functions, with b-period equal to

$$B_{11} = 2 \int_c^a dv_1(\lambda) \, , \tag{3.4.3}$$

where

$$dv_1 = N \frac{d\lambda}{\sqrt{P(\lambda)}} \, .$$

Here,

$$P(\lambda) = (\lambda - a)(\lambda - \bar{a})(\lambda - c)(\lambda - \bar{c}) \, ,$$

$$N = \left[2 \int_0^\infty \frac{d\lambda}{\sqrt{P(\lambda)}} \right]^{-1} \, . \tag{3.4.4}$$

In (3.4.4) the integration is carried out over γ_1, indicated in Fig. 3.5. In (3.4.2) we have:

$$\alpha_\mp = -2iN[x - \text{Re}(c + a)t] - \eta_1 \mp \Gamma_1 \, . \tag{3.4.5}$$

Here,

$$\eta_1 = d - i(\text{Im}\,\Gamma_1) \, ,$$

$$\Gamma_1 = \int_c^\infty \sum dv_1(\lambda) \, ,$$

where d is an arbitrary real number and integration is over γ_2.

The branch of $\sqrt{P(\lambda)}$ is assumed to be chosen so that $\sqrt{P(\lambda)} > 0$, $\lambda \to \infty$, $\lambda \in \mathcal{R}$. It is assumed in (3.4.3) that

$$B_{21} = 2 \int_{\bar{c}}^\lambda dv_1(\lambda) \, ,$$

$$\eta_+ - \eta_- = -2\Gamma_2 \, .$$

where

$$\Gamma_2 = \frac{1}{(2\pi i)^2} \frac{\sqrt{P(\lambda_0)}}{(\lambda + \bar{\lambda}_0 + \Lambda_0)} \int_a^\infty \frac{\lambda + \lambda_0 + \Lambda_0}{(\lambda - \lambda_0)\sqrt{P(\lambda)}} d\lambda \, , \tag{3.4.6}$$

$$V = -\frac{\sqrt{P(\lambda_0)}}{2(\lambda_0 + \bar{\lambda}_0 + \Lambda_0)} \, ,$$

$$W = -\frac{\sqrt{P(\lambda_0)}}{4(\lambda_0 + \bar{\lambda}_0 + \Lambda_0)} \{ \text{Re}(\lambda_0 + a + c) + \Lambda_0 \} \, .$$

In the latter formulae

$$\Lambda_0 \equiv -\int_{\bar{a}}^a \frac{(\lambda + \lambda_0)d\lambda}{(\lambda - \lambda_0)\sqrt{P(\lambda)}} \left[\int_{\bar{a}}^a \frac{d\lambda}{\sqrt{P(\lambda)}} \right]^{-1} \, .$$

Further, we have

$$B_{32} = \frac{\sqrt{P(\lambda_0)}}{\pi i(\lambda_0 + \bar{\lambda}_0 + \Lambda)} \int_{\bar{c}}^{\bar{\lambda}_0} \frac{(\lambda + \bar{\lambda}_0 - \Lambda_0)d\lambda}{(\lambda - \bar{\lambda}_0)\sqrt{P(\lambda)}} .$$ (3.4.7)

It is necessary to determine the quantitites A_0, E_0, N_0, entering in (3.4.1). Let us consider the integrals

$$\Omega_1(\lambda) = \int_a^{\lambda} \frac{\lambda^2 + a_0}{\sqrt{P(\lambda)}} \, d\lambda ,$$

where

$$a_0 = - \int_{\bar{a}}^{a} \frac{\lambda^2 d\lambda}{\sqrt{P(\Lambda)}} \left[\int_{\bar{a}}^{a} \frac{d\lambda}{\sqrt{P(\lambda)}} \right]^{-1} ,$$

$$\Omega_2(\lambda) = \int_a^{\lambda} \frac{2\lambda^3 + b_0}{\sqrt{P(\lambda)}} \, d\lambda ,$$

where

$$b_0 = \int_{\bar{a}}^{a} \frac{2\lambda^3 d\lambda}{\sqrt{P(\lambda)}} \left[\int_{\bar{a}}^{a} \frac{d\lambda}{\sqrt{P(\lambda)}} \right]^{-1} ,$$

$$\Omega_3 = \int \frac{-\lambda + c_0}{\sqrt{P(\lambda)}} \, d\lambda ,$$

where

$$c_0 = \int_{\bar{a}}^{a} \frac{\lambda d\lambda}{\sqrt{P(\lambda)}} \left[\int_{\bar{a}}^{a} \frac{d\lambda}{\sqrt{P(\Lambda)}} \right]^{-1} .$$

As assumed previously, $\sqrt{P(\lambda)} > 0$, $\lambda \to \infty$, $\lambda \in \mathcal{R}$. It is clear that

$$\Omega_1(\lambda) = \lambda + A + O(1/\lambda) ,$$

$$\Omega_2(\lambda) = 2\lambda^2 + B + O(1/\lambda) ,$$

$$\Omega_3(\lambda) = -\ln \lambda - c + O(1/\lambda) .$$

By definition,

$$E_0 = 2A , \quad N_0 = 2B , \quad A_0 = c .$$ (3.4.8)

Let us analyze the solution obtained. As from (3.4.1,2) the solution can be interpreted as the interaction of a cnoidal wave travelling with the velocity $V = \mathrm{Re}(c + a)$ (roughly speaking, the function $\theta_3(\lambda)$ is responsible for the cnoidal wave) with the solution propagating with the velocity $\hat{\lambda}_0 = \mathrm{Im}\, W / \mathrm{Im}\, V$:

$$\hat{\lambda}_0 = 2\mathrm{Re}(\lambda_0 + a + c + \Lambda_0) - (\mathrm{Im}\, S)/(\mathrm{Re}\, S)\mathrm{Im}\, \Lambda_0 ,$$ (3.4.9)

where

$$S = \frac{1}{2\pi i} \frac{\sqrt{P(\lambda)}}{(\lambda_0 + \bar{\lambda}_0 + \Lambda_0)} .$$

Now, we will briefly analyze a general solution of (3.4.2).

In the absence of a cnoidal wave the solution travels with the velocity $V_0 = 4\mathrm{Re}\,\lambda_0$. Taking into account the interaction with the periodic solution, one can see that the soliton velocity changes. The effect of a soliton on a cnoidal wave is significant only in a narrow region near a "soliton beam" $\beta = 0$, where

$$\beta = (\mathrm{Im}\, V)x - (\mathrm{Im}\, W)t . \tag{3.4.10}$$

In the regions $\beta << 0$ and $\beta >> 0$ the solution turns into an ordinary cnoidal wave. In particular, for $\beta << 0$ the wave has the form

$$q_1(x,t) = \frac{\theta_3(\alpha_- - 2\mathrm{Re}\, B_{21})}{\theta_3(\alpha_+ - 2\mathrm{Re}\, B_{21})} \exp[i(E_0 x + N_0 t) + A_0] , \tag{3.4.11}$$

and for $\beta >> 0$:

$$q_2(x,t) = \frac{\theta_3(\alpha_-)}{\theta_3(\alpha_+)} \exp[i(E_0 x + N_0 t) + A_0] , \tag{3.4.12}$$

So, the effect of the soliton solution on the cnoidal wave consists of a phase shift of the latter by the value

$$\Delta = 2\mathrm{Re}\, B_{21} = 4 \int_c^{\lambda_0} \frac{N d\lambda}{\sqrt{P(\lambda)}} . \tag{3.4.13}$$

For some λ_0 the phase shift Δ is zero. This is a consequence of the fact that $\int_c^{\lambda_0} dv_1(\lambda)$ has the whole plane C as the region of its values (a particular case of a known theorem that the Abelian mapping $\sum_j^n \int_c^{\lambda} dV$ is surjective onto the Jacobian $J_n(\Gamma)$ of a compact Riemann surface Γ). In this case the structure of solution (3.4.1) is greatly simplified. The solution is a product of two factors

$$q(x,t) = q_2(x,t)\mathcal{E}(x,t) . \tag{3.4.14}$$

Here, $q_2(x,t)$ is given by (3.4.12) which describes an ordinary cnoidal wave; $\mathcal{E}(x,t)$ is a perturbation caused by a soliton:

$$\mathcal{E}(x,t) = \frac{1 + e^{\xi_-} + e^{\bar{\xi}_-} + e^{\xi_- + \bar{\xi}_- + 2\pi i B_{32}}}{1 + e^{\xi_+} + e^{\bar{\xi}_+} + e^{(\xi_+ + \bar{\xi}_+) - 2\pi i B_{32}}} , \tag{3.4.15}$$

where

$$\xi_\pm = -i(V x - W t) + \eta_\pm ,$$

W, V, $\eta_\pm$ are defined in (3.4.6,7).

With the notation of (3.4.15), formula (3.4.2) is greatly simplified:

$$\theta_{\pm}(x,t) = \theta_3(\alpha_{\pm}) + \theta_3(\alpha_{\pm} - B_{21})e^{\xi_{\pm}} + \theta_3(\alpha_{\pm} - \bar{B}_{21})e^{\bar{\xi}_{\pm}}$$
$$+ \theta_3(\alpha_{\pm} - 2\mathrm{Re}\, B_{21})e^{(\xi_{\pm} + \bar{\xi}_{\pm} - 2\pi i B_{32})} . \qquad (3.4.16)$$

Equations (3.4.1,12) give a complete description of the interaction between a soliton and a cnoidal wave.

3.5 Solitons in Two Tunnel-Coupled Optical Waveguides

The investigation of soliton interaction in optical waveguides is very important for the development of information transmission systems based on solitons. The interaction between solitons imposes some restrictions on the rate of information transfer, so we need some special procedures to suppress the interaction. On the other hand, the soliton interaction can be used to form soliton trains which can be used as part of the transmission system.

New perspectives appear when we investigate soliton interaction in a system of two tunnel-coupled optical waveguides. Such systems have attracted some attention in problems dealing with the evolution of continuous pump waves. *Jensen* (1982) and *Mayer* (1982, 1985) demonstrated that, using two coupled fibres, one can create an optical switch, an optical transistor, and other simple logical devices.

The numerical calculations performed by *Andrushko* et al. (1987) and *Trillo* et al. (1988) showed that in this system, soliton switching is possible when a soliton propagates along one of the fibres.

The problem of soliton interaction in two coupled fibres is considered analytically and numerically. Conditions for the existence of a coupled soliton state are given, and criteria for its decay are obtained.

In standard dimensionless variables, the equations for the envelopes of the electric fields in two identical fibres are given by the coupled nonlinear Schrödinger equations,

$$i q_{n\xi} + \frac{1}{2} q_{n\tau\tau} + |q_n|^2 q_n = \varepsilon q_m ,$$

$$n, m = 1, 2 \quad (n \neq m) . \qquad (3.5.1)$$

Here, ξ is the normalized distance of propagation, τ is the normalized retarded time, and ε is the coefficient of coupling between the fibres, which can be calculated by using a Gaussian approximation for the field mode

$$\varepsilon \simeq \frac{\Delta n}{n_2 |E_0|^2} \exp\left(-\frac{\ell^2}{4\varrho_0^2}\right) .$$

In this expression, Δn is the difference between the indices of refraction of the fibres and the supporting medium, ℓ is the distance between the fibres, ϱ_0 is the modal spot size, and E_0 is the initial amplitude of the electric field in the fibres. In the soliton regime, the influence of a weak wave in one fibre on a soliton propagating in the neighbouring fibre has been investigated analytically (Sect. 2.7). The propagation of a single soliton pulse in two coupled fibres has also been studied numerically.

Now, we investigate the system (3.5.1) for the case when the initial state in each fibre is a soliton (*Abdullaev* et al., 1989; *Heatley* et al., 1988; *Wright* et al., 1989), so that

$$q_n = 2\eta_n \text{sech} \left[2\eta(\tau - \zeta_n) \right] \exp \left[2i\mu_n(\tau - \zeta_n) + i\delta_n \right] , \tag{3.5.2}$$

$$n = 1, 2 ,$$

where η_n, μ_n, δ_n, ζ_n are the parameters characterizing the amplitude, velocity, phase, and position of the centre of the soliton, respectively. If ε is small, i.e., in the weak-coupling limit, we can use a perturbation theory for solitons based on the IST to find a solution. For (3.5.1), the perturbation operator takes the form

$$R_{mn} = -iq_m , \quad n, m = 1, 2 \quad (n \neq m) .$$

Now, we introduce new variables:

$$\eta = \frac{1}{2}(\eta_1 + \eta_2) , \quad r = 2\eta(\zeta_2 - \zeta_1) , \quad \psi = \delta_2 - \delta_1 ,$$

$$\Delta\mu = \mu_2 - \mu_1 , \quad \Delta\zeta = \zeta_2 - \zeta_1 .$$

In the adiabatic approximation, the following equations for the soliton parameters are obtained:

$$\mu_{n\xi} = (-1)^n 2\eta\varepsilon \frac{r \cosh(r) - \sinh(r)}{\sinh^2(r)} \cos(\psi) , \tag{3.5.3}$$

$$\eta_{n\xi} = (-1)^n 2\eta\varepsilon \frac{r}{\sinh(r)} \sin(\psi) , \tag{3.5.4}$$

$$\zeta_{n\xi} = 2\mu_n - \frac{\varepsilon}{2\eta_n} \frac{r^2}{\sinh(r)} \sin(\psi) , \tag{3.5.5}$$

$$\delta_{n\xi} = 2\mu_n \zeta_{n\xi} + 2(\eta_n^2 - \mu_n^2)$$

$$+ \varepsilon \frac{r}{\sinh(r)} \left(\frac{r}{\tanh(r)} - 2 \right) \cos(\psi) . \tag{3.5.6}$$

In deriving these, we have assumed that $\mu_{n0} = 0$, $\eta_{10} = \eta_{20}$, and $\eta_1 \approx \eta_2 = \eta$. As seen from (3.5.4,5), the assumption of constant amplitude for the interacting solitons holds when the initial phase difference $\psi_0 = \delta_{20} - \delta_{10} = 0$; π.

We next analyze the system of equations (3.5.3) and (3.5.6). From (3.5.3) and (3.5.4), we can easily show that the following integrals of motion exist

$$\mu = \frac{1}{2}(\mu_1 + \mu_2) = 0 , \tag{3.5.7}$$

$$\zeta_1 + \zeta_2 = 0 , \tag{3.5.8}$$

$$\psi = \psi_0 \ . \tag{3.5.9}$$

Then we find that

$$\Delta\mu_\xi = -4\eta\varepsilon \ \frac{r\cosh(r) - \sinh(r)}{\sinh^2(r)} \ \cos(\psi_0) \ , \tag{3.5.10}$$

$$\Delta\zeta_\xi = 2\Delta\mu \ . \tag{3.5.11}$$

Differentiating (3.5.11) with respect to ξ and taking into account (3.5.10), we obtain a closed equation for r,

$$r_{\xi\xi} - 16\eta^2\varepsilon\cos(\psi_0) \ \frac{r\cosh(r) - \sinh(r)}{\sinh^2(r)} = 0 \ . \tag{3.5.12}$$

It follows from (3.5.12) that the potential for the interaction between the solitons is

$$U = 16\eta^2\varepsilon \ \frac{r}{\sinh(r)} \cos(\psi_0) \ . \tag{3.5.13}$$

From (3.5.13), and from an analysis of (3.5.12) in the phase plane, it follows that there exists a bound state for the solitons when $\varepsilon\cos\psi_0 < 0$, i.e., $\psi_0 = \pi$ at $\varepsilon > 0$ (or $\psi_0 = 0$ at $\varepsilon < 0$). For small r ($r \ll 1$), (3.5.12) reduces to a simple harmonic oscillator with a period of oscillation $L = \pi/(2\eta\sqrt{|\varepsilon|/3})$. For $\psi_0 = 0$, $\varepsilon > 0$, the potential of interaction becomes the potential of repulsion. For $\psi_0 \neq 0, \pi$, substantial energy exchange between solitons occurs, and the analysis developed here cannot be applied.

This system of equations was also solved numerically. The numerical experiments were carried out for those system parameters for which analytical results were available, and also for those values where an analytical approach is not possible. All calculations were made for the case of equal initial soliton amplitudes ($2\eta_{10} = 2\eta_{20} = 1$). Under the weak-coupling condition ($0 < \varepsilon \ll 1$), the numerical experiments were in agreement with analytical calculations. Both the attractive state for the solitons when $\psi_0 = \pi$, as well as the repulsive state when $\psi_0 = 0$, were observed (Fig. 3.6). There is also a coupled state for strong coupling ($\varepsilon = 1$) when $\psi_0 = \pi$, but in this case the pulse amplitude is periodically changed and takes a maximum value when the distance between the solitons is a maximum. In-phase solitons evolve in such a way that their soliton-like form changes (Fig. 3.7). We stress that when the phase difference ψ_0 equals 0, or π, a redistribution of energy between the interacting pulses does not occur (the energy in each fibre is conserved). Such special behaviour does not occur when $0 < \psi_0 < \pi$; then the major portion of pulse energy can be transferred from one waveguide to another (Fig. 3.8). In addition to energy transfer, we note a state of coupled pulses for $\pi/2 < \psi_0 < \pi$.

In summary, we have determined conditions under which solitons form a coupled state, and conditions when a repulsive force arises, causing the solitons to separate during propagation. We have also shown that, with an appropriate choice of soliton phase difference, energy transfer between fibres is possible. Such dynamics affords a wide variety of opportunities to develop new elements and devices useful in soliton-based OFC systems.

Recent developments that supplement the exposition of this Chapter refer to a study of *Friberg* (1991) into the fusion and steering of two different-color solitons, to the investigation of *Friberg* and *DeLong* (1992) into the influence of a perturbation of the bound state of solitons on break up. Essential progress has been achieved in soliton coupler theory and consideration has been given to the switching of SIT (self-induced transparency) solitons in a two-level system – doped NLDS – by *Guzman* et al. (1990), *Romagnoli* et al. (1991), *Abdullaev* and *Gulyamov* (1992). The influence of SRS and amplification on NLDC was studied by *Wabnitz* et al. (1991) and *Abdullaev* et al. (1992).

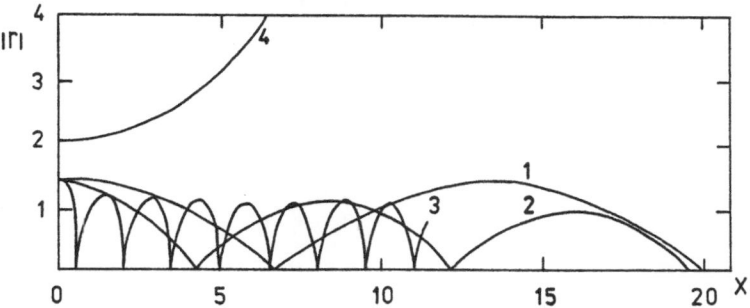

Fig. 3.6. Variation of the interval between solitons $r = 2\nu(\zeta_2 - \zeta_1)$ with the distance along the fibres for different initial phases ψ_0 and coupling coefficients ε. Curve *1*: $\varepsilon = 0.05$ and $\psi_0 = \pi$; curve 2: $\varepsilon = 0.1$ and $\psi_0 = \pi$; curve *3*: $\varepsilon = 1.0$ and $\psi_0 = \pi$; curve 4: $\varepsilon = 0.05$ and $\psi_0 = 0$

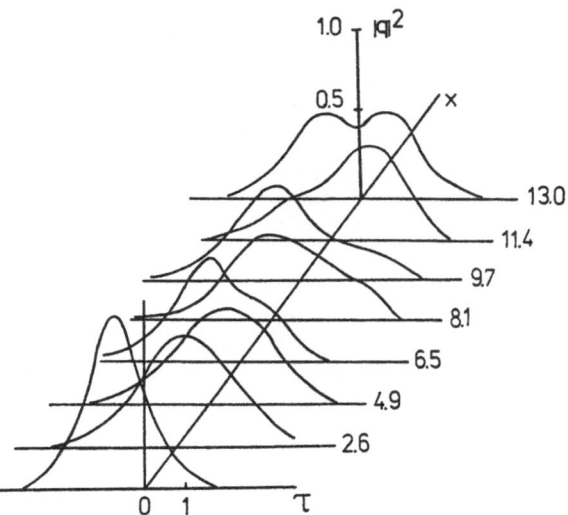

Fig. 3.7. Deformation of a soliton for $\varepsilon = 1.0$ and $\psi_0 = 0$ along the length of a fibre

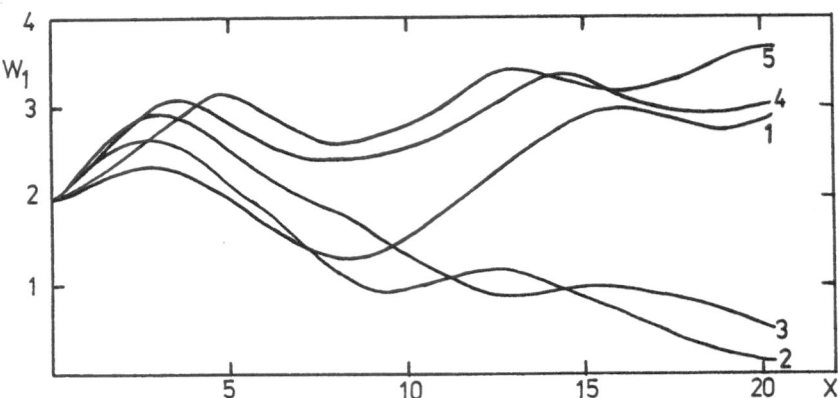

Fig. 3.8. Energy transfer from one fibre to another when $0 < \psi < \pi$ ($\varepsilon = 0$). $W_1 = \int |q_1|^2 d\tau$ is the energy of the pulse in a fibre ($W_1 + W_2 = $ const). Curve *1*: $\psi_0 = \pi/6$; curve *2*: $\psi_0 = \pi/3$; curve *3*: $\psi_0 = \pi/2$; curve *4*: $\psi_0 = 2\pi/3$; curve *5*: $\psi_0 = 5\pi/6$

4. Statistical Dynamics of Optical Solitons

In analysis of pulse propagation in optical fibres, stochastic effects must be taken into account. These effects can be classified into three basic types:

1) stochasticity associated with the chaotic nature of the initial pulse, due to partial coherence of the laser-generated radiation;
2) stochasticity due to random nonuniformities in the optical fibre (fluctuations in the values of the dielectric constant, of the fibre diameter, and so on);
3) the chaotic field caused by a "dynamical stochasticity", such as might arise from a periodic modulation of the system parameters, or when a periodic array of pulses propagates in a fibre optic resonator.

Methods for analysing each class of effects are different. It is worth noting that when the system is described by integrable nonlinear wave equations, or nearly integrable ones, problems of evolution of random light fields can be solved by the IST. Elsewhere, asymptotic methods (direct methods) should be applied.

4.1 Random Pulses in an Optical Fibre.
Numerical Simulation in the Presoliton Region

Here, we will describe results of a numerical simulation of the evolution of pulses with random amplitude and phase modulation (*Akhmanov* et al., 1986). Let the initial pulse be of the form

$$q(\tau, 0) = q_0(\tau)(1 + \sigma_\varepsilon \varepsilon(\tau)) \ ,$$

$$q_0(\tau) = A \exp\left(-\frac{\tau^2}{2\tau_0^2}\right) \ , \tag{4.1.1}$$

where $\varepsilon(\tau) = \varepsilon_R + i\varepsilon_I$ is a random complex Gaussian process with a zero mean value and unit dispersion; σ_ε characterizes the noise level. The correlation function is

$$\langle \varepsilon(t)\varepsilon^*(t + \theta) \rangle = \exp\left(-\frac{\theta^2}{\tau_{co}^2}\right) \ . \tag{4.1.2}$$

The angular brackets $\langle \ldots \rangle$ indicate an average over the ensemble ε. Numerical data for the evolution of intensity and frequency were obtained by the Monte-Carlo method. In Fig. 4.1 they are given as functions of the distance z. As shown, intensity fluctuations become smoothed, and frequency modulation in the central part of the pulse is linearized. Fig. 4.1 illustrates temporal intensity distribution and normalized addition to the carrier frequency of partially-coherent pulse. Here, $\sigma = 0.2$; $\tau_c/\tau_0 = 0.64$, $R = 300$, $z = 2z_d = 2(kn_2I)^{-1}$, $R = z_d/z_{nl}$, $z_d = \tau_0^2/k''$; τ_0 is the initial pulse duration. The numerical analysis shows that the intensity fluctuations are smoothed, but the dependence of the current frequency value on time (sweep) is linearized. At optimal for compression distances $z \sim 2(z_d z_{nl})^{1/2}$ the dispersion due to the frequency spread of the velocities (corresponding to various realizations of random function) is reduced. The fluctuations, in general, influence temporal envelope fronts; the high-frequency and low-frequency wings of a spectrum correspond to fronts.

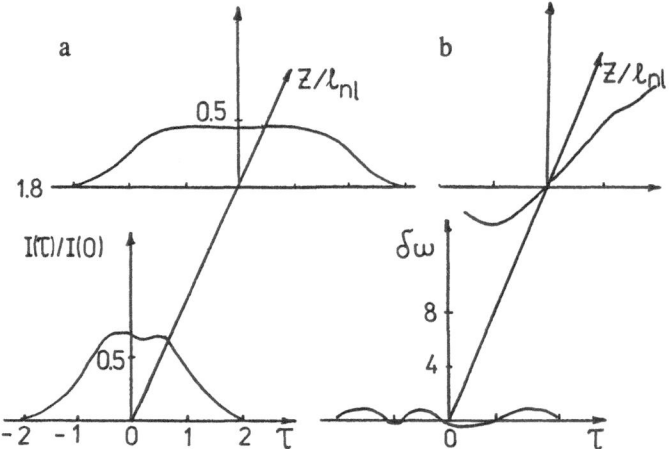

Fig. 4.1. Shape variation of random pulse (**a**), and distribution of instantaneous frequency (**b**) over the length z of an optical fibre. The lower plots show the initial shape and frequency distribution, the upper plots show the data for the same pulse at $z/l_{nl} = 1.8$

Vysloukh and *Fattakhov* (1986) have developed a theory to describe these results. They have applied a method of momentum and found an evolution of the mean square duration of the statistically averaged pulse. The mean square pulse duration is defined by the relationship

$$\theta(\zeta) = \int_{-\infty}^{\infty} q(\tau)\tau^2 q^*(\tau)d\tau .$$

The mean square value of duration averaged over the ensemble of random initial data, $\langle \theta \rangle$ is represented by superposition of regular $\bar{\theta}_0$ and random $\tilde{\theta}$ components:

$$\langle \theta(\zeta) \rangle = \bar{\theta}_0(\zeta) + \sigma^2 \tilde{\theta}(\zeta) .$$

Using the NLS equation and these two equations, one can show that the initial stage of pulse evolution ($\zeta \ll 1$) is described by the system:

$$\frac{d^2\bar{\theta}_0}{d\zeta^2} = \sqrt{\pi}\left(1 + \frac{R}{\sqrt{2}}\right) ,$$

$$\frac{d^2\tilde{\theta}}{d\zeta^2} = \sqrt{\pi}\left[1 + 4\left(\tau_c^{-2} + \frac{R}{\sqrt{2}}\right)\right] .$$

It was assumed that the initial conditions have been taken in the form $q_0(\tau, 0) = \exp(-\tau^2/2\tau_c^2)$, $\tilde{q}(\tau, 0) = \xi(\tau)q_0$. As it follows from this system of equations, θ_0 and $\tilde{\theta}$ both increase proportionally to the square of the distance. When $R \gg 1$, the velocity of a noise-component-broadening is four times faster than a regular one. Moreover, the intensity of fluctuations is localized in the high- and low-frequency parts of the spectrum, which allows control of pulse parameters.

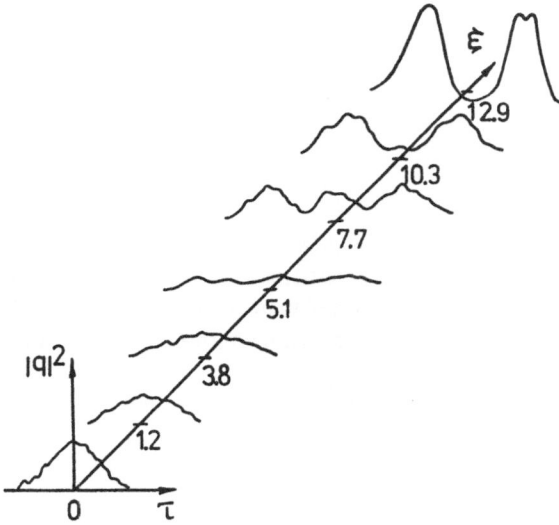

Fig. 4.2. Numerically calculated evolution of a pulse for a gain factor $\Gamma = 0.15$

Of much interest is the problem of self-action of randomly modulated pulses in an active fibre when the gain is distributed uniformly over the whole length of the fibre. The corresponding equation has the form of a modified NLS equation:

$$iq_\xi + q_{\tau\tau} + 2|q|^2 q = i\Gamma q . \tag{4.1.3}$$

Equation (4.1.3) describes pulse evolution in the optical fibre-amplifier system (*Azimov* et al., 1986).

Figure 4.2 illustrates results of a relevant numerical simulation (*Khikmatov*, 1987). Plots of the pulse intensity $|q|^2$ are shown for different values of propagation distance. Parameter values are $A = 0.7$; $\tau_0 = 0.19$; $\Gamma = 0.15$; while h is a discretization value of the coordinate. For small values of the propagating distance $\xi \sim 1.5$, fluctuations of the pulse are smoothed and the pulse amplitude decreases. For $\xi \sim 5$, the pulse broadens, for $\xi = 8$ three subpulses have

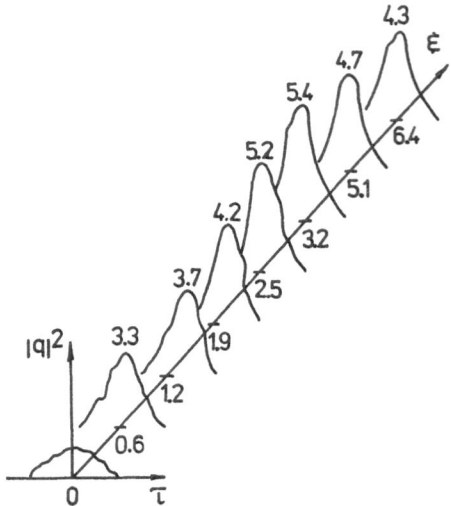

Fig. 4.3. As in Fig. 4.2 for $\Gamma = 3$

formed, and for $\xi \sim 13$ two $N = 2$ solitons have been generated. Evolution of initial pulses for large values of gain $\Gamma = 3$ have also been studied. It is found that phase and amplitude fluctuations decrease faster than for the case $\Gamma \ll 1$ (Fig. 4.3). Fluctuations are smoothed at $\xi \sim 0.1$, when the amplitude grows until, at $\xi \sim 6.4$, and $N = 2$ soliton forms.

The accuracy of the simulation has been checked by observing the change in the integral invariant n which, according to theory, must satisfy

$$N = N_0 \exp(2\Gamma\xi) .$$

The agreement of the numerical data with this formula is quite good (Fig. 4.4).

To sum up, a conclusion can be made that uniformly distributed amplification along the fibre is useful, not only in compensating for pulse energy loss, but also for suppressing weak random signal distortions.

In the next section we will analyze the propagation of pulses which are initially stochastic in the fibre (presolitonic region) using the path integral method. We will also investigate the propagation of a small random wave packet in the fibre under the possible amplification. For this purpose we will apply an appropriate perturbation theory based on the IST for the NLS equation in the presoliton region.

It is convenient to go from the NLS equation over to canonical action-angle variables $n(\lambda)$, ϕ (Appendix B). Applying the IST, we find for $n(\lambda)$:

$$n(\lambda, \xi) = n(\lambda, \ \xi = \xi_0) \exp(2\Gamma\xi) . \qquad (4.1.4)$$

The solution in τ-space has the form

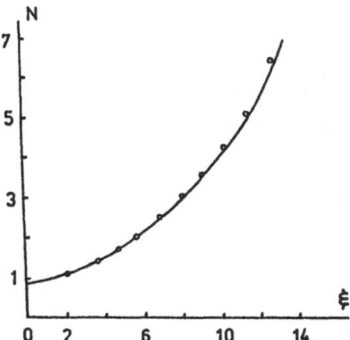

Fig. 4.4. ξ-dependence of N from data of Figs. 4.2, 3. Numerical results are denoted by dots, theoretical data by the solid line

$$|q|^2 \simeq \frac{1}{4\xi}\, n\left(\lambda = -\frac{\tau}{4\xi}\right) = \frac{1}{4\xi}\, n_0\left(\lambda = -\frac{\tau}{4\xi_0}\right) \exp(2\Gamma\xi)\,, \qquad (4.1.5)$$

$$\Phi(\tau,\xi) = \arg q(\tau,\xi) = \frac{\tau^2}{4\xi} + \frac{1}{2}\, n\left(\lambda = -\frac{\tau}{4\xi}\right) \ln(\xi\tau_0^{-2})\,. \qquad (4.1.6)$$

The $n(\lambda)$ is connected with the Jost coefficient of the Zakharov-Shabat linear spectral problem $a(\lambda)$ by

$$n(\lambda) = \ln |a(\lambda)|^2\,,$$

(see Appendix B). Taking $|a|^2 = 1 + |b|^2$ into account, one obtains for $|b|^2 \ll 1$

$$n_0 \approx \frac{1}{2}\, |b|^2\,,$$

and we derive a mean value,

$$\langle n_0\rangle \approx \frac{1}{2}\, \sigma_\varepsilon^2 \tau\,.$$

Substitution of n_0 into (4.1.5,6) yields

$$\langle |q|^2\rangle \approx \left[\frac{\sigma_\varepsilon^2 \tau_0}{8\xi}\right] \exp(2\Gamma\xi)\,,$$

$$\Phi = \frac{\tau^2}{4\xi} + \sigma_\varepsilon^2 \tau_0 \exp(2\Gamma\xi)\frac{\ln(\xi/\tau_0^2)}{8\xi}\,. \qquad (4.1.7)$$

These expressions indicate that during the initial stages of pulse propagation in the fibre, its amplitude should decrease and its profile should become smoother; then, as ξ increases the amplitude should increase. This is observed in numerical simulations (Fig. 4.2).

4.2 Noise Signals in the Near Field

A number of analytic results can be obtained by considering the evolution of random pulses while propagating over short distances in a fibre, when $\xi << 1$ (*Fattakhov* and *Chirkin*, 1983).

We will treat $|q(\xi, \tau)|^2$ as an initial potential from which a pulse propagates. Then, a solution to the NLS equation can be found in the form

$$q(\xi, \tau) = \int_{-\infty}^{\infty} q_0(\theta) G(\theta, \tau; \xi) d\theta , \qquad (4.2.1)$$

where $q_0(\theta) = q(\theta, \xi = 0)$ is the initial condition; $G(\theta, \tau; \xi)$ is a Green's function that can be expressed in terms of the path integral

$$G(\theta, \tau; \xi) = \int \exp \left\{ - \int_0^{\zeta} L\left[\tau(x), \dot{\tau}(x)\right] dx \right\} D\tau(x) , \quad \tau(0) = 0 ,$$

$$\dot{\tau} = \frac{d\tau(x)}{dx} , \qquad (4.2.1a)$$

$$L = \frac{1}{2} \dot{\tau}^2 + R|q|^2 , \quad R = \frac{z_d}{z_{nl}} .$$

Here L is the Langrange function. The maximum contribution to this integral arises from paths which satisfy the Euler equations

$$\frac{\partial}{\partial x} \frac{\partial L}{\partial \dot{\tau}} - \frac{\partial L}{\partial \tau} = 0 ,$$

i.e., in our case,

$$\ddot{\tau} - R \frac{\partial}{\partial \tau} |q(\tau(x), x)|^2 = 0 . \qquad (4.2.1b)$$

For $\xi < 1$, we can solve the NLS equation iteratively. In the lowest order approximation, one can neglect the dispersion term in the NLS equation. Then,

$$q^{(0)}(\tau, \xi) = q_0(\tau) \exp \left[i|q_0(\tau)|^2 \xi\right] . \qquad (4.2.2)$$

Using (4.2.2) and the path integration method with $|q_0|^2 = \exp(-\tau^2)$, we obtain from (4.2.1b) at $\tau < 1$ the solution

$$\tau(x) = \theta \cos(h_0 x) + \sin(h_0 x)[\tau - \theta \cos(h_0 x)] \sin^{-1}(h_0 \xi) ,$$

$$h_0 = (2R)^{1/2} .$$

Substituting this expression into (4.2.1a), we obtain an approximation for the Green's function $G(\theta, \tau; \xi)$ (*Akhmanov* et al., 1986)

$$G(\theta, \tau, \xi) = \left[-\sqrt{2}\pi i \sin \sqrt{2}\xi \right]^{-1/2} \exp\left\{ \frac{i\sqrt{2}}{\sin(\sqrt{2}\xi)} \right.$$

$$\left. \times \left[(\theta^2 + \tau^2) \cos \sqrt{2}\xi - 2\tau\theta \right] + i\xi \right\} . \tag{4.2.3}$$

Let us now consider the effects of pulse self-action with a random phase, with

$$q_0(\tau) = \exp\left[-\frac{\tau^2}{2} + i\phi_0(\tau) \right] . \tag{4.2.4}$$

Statistical properties of the phase $\phi_0(\tau)$ are defined by

$$\langle \phi_0 \rangle = 0 , \quad \langle \phi_0(\tau_1)\phi_0(\tau_2) \rangle = \sigma_\phi^2 \exp\left[-\frac{(\tau_1 - \tau_2)^2}{\tau_\phi^2} \right] .$$

Here τ_ϕ is the correlation time of phase fluctuations.

With the aid of (4.2.1,4), a correlation function for a pulse with random initial modulation propagating over a distance in a nonlinear fibre is obtained as

$$B(\tau_2, \tau_1; \xi) = \langle q(\tau_1, \xi) q^*(\tau_2, \xi) \rangle$$

$$= \frac{1}{V_2(\xi)} \exp\left\{ -\frac{1}{2V_2^2(\xi)} \left[\tau_1^2 + \tau_2^2 - 2\left(\frac{\sigma_\phi}{\tau_\phi} \right)^2 \right. \right.$$

$$\left. \left. \times (\tau_2 - \tau_1)^2 \right] \right\} . \tag{4.2.5}$$

Here,

$$\tau_\phi = \frac{\tau_c}{\tau_0} ,$$

$$V_2^2(\xi) = 1 + \left\{ \left[\frac{1}{4} + \left(\frac{\sigma_\phi}{\tau_\phi} \right)^2 \right] - 1 \right\} \sin^2(\sqrt{2}\xi) .$$

An ensemble-averaged value for the pulse duration τ_u is seen to be oscillating according to the law

$$\tau_u(\xi) = \tau_0 \left\{ 1 + \left[\frac{1}{4} + \left(\frac{\sigma_\phi}{\tau_\phi} \right)^2 - 1 \right] \sin^2 \sqrt{2}\xi \right\}^{1/2} .$$

Note that for some values of the parameters, the ensemble-averaged pulse propagates with constant duration.

Later on, we will examine the self-action of a random pulse of the form

$$q_0(\tau) = \varepsilon(\tau) F(\tau) ,$$

where $\varepsilon(\tau)$ is a random Gaussian process, and $F(\tau) = \exp(-\tau^2/2)$. With the approximation (4.2.3), the correlation function has the value

$$B(\tau_2, \tau_1; \xi) = \iint_{-\infty}^{\infty} d\theta_1 \, d\theta_2 \iint_{-\infty}^{\infty} K(\theta_1, \theta_2; \tau_1 . \tau_2)$$

$$\times \exp\left\{-\frac{i}{2} \int_0^{\xi} \left(\dot{\tau}_1^2(x) - \dot{\tau}_2^2(x)\right) dx\right\} D\tau_1 D\tau_2 \,,$$

where the "correlator" K has the form

$$K(\theta_1, \theta_2; \tau_1, \tau_2) = \left\langle q_0(\theta_1) q^*(\theta_2) \exp\left\{-i \int_0^{\xi} \left[|q_0(\tau_1)|^2 - |q_0(\tau_2)|^2\right] dx\right\}\right\rangle.$$

If the propagation is coherent, i.e., $\xi < L_{\text{coh}} = \tau_{\text{co}}^2/|k''|$, then for averaging, the replacement

$$\int_0^{\xi} |q_0(\tau(x))|^2 dx \approx |q_0(\tau(0))|^2 \xi = |q_0(\theta)|^2 \xi \,,$$

can be used. A simple calculation then shows that the pulse length is described by the expression

$$\tau_p(\xi) = V_2(\xi)\tau_0 \,.$$

The results obtained by *Fattakhov* and *Chirkin* (1984) show that for the incoherent mode of propagation, where $\xi > L_{\text{coh}}$, the mean pulse length and the correlation time always grow, whereas for the coherent mode they can either decrease or increase. Numerical modelling by *Kandidov* and *Schlemov* (1986) confirm these results.

4.3 Noise Signals in the Far Field

It is convenient to use the IST to describe the evolution of a partially coherent pulse in the asymptotic region ($\xi \geq z_{nl}$). The method has been applied to describe nonlinear Fraunhofer diffraction of random fields in Kerr-type nonlinear media, and to the propagation of nonlinear incoherent pulses in fibres (*Bass* et al., 1986,1988).

The method for solving this problem can be formulated as follows:

1) Starting with an initial random pulse, statistics of spectral data for the Zakharov-Shabat system are found.
2) The solutions of the NLS equation are restored by the aid of the IST from the scattering data. A random initial condition evolves into the solitons plus quasi-linear oscillating wave packets. The parameters of formed solitons and plane waves are random. The corresponding distribution function for the parameters is restored with the help of the inverse scattering transform.
3) Mean characteristics and appropriate correlation functions for the transversed pulse are then determined.

Thus, in the first step, we start with a definition scattering data for the Zakharov-Shabat system:

$$\frac{\partial\psi^{(1)}}{\partial\tau} = i\lambda\psi^{(1)} + iq_0^*(\tau)\psi^{(2)} ,$$

$$\frac{\partial\psi^{(2)}}{\partial\tau} = -i\lambda\psi^{(2)} + iq_0(\tau)\psi^{(1)} ,\qquad (4.3.1)$$

where $\psi = (\psi^{(1)}, \psi^{(2)})$, and λ is a spectral parameter.

The solution of the spectral problem is represented by scattering data for discrete and continuous spectra. To obtain this data, one has to know the Jost functions. Let the initial potential have the form

$$q(\tau,0) = \begin{cases} 0 , & \tau < 0 , \quad \tau > T , \\ q_0(\tau) + \varepsilon(\tau) , & 0 < \tau < T . \end{cases}\qquad (4.3.2)$$

For $t = 0$, the corresponding Jost functions are

$$\psi(\tau,\lambda) = \begin{cases} = \begin{Bmatrix} a(\lambda)\exp(i\lambda\tau) \\ b^*(\lambda)\exp(-i\lambda\tau) \end{Bmatrix} ,\tau < 0 \\[2ex] = e^{i\lambda T}\begin{Bmatrix} \cos[\chi(\tau - T)] + \frac{i\lambda}{\chi}\sin[\chi(\tau - T)] \\ \frac{iq_0}{\chi}\sin[\chi(\tau - T)] \end{Bmatrix} ,0 < \tau < T \\[2ex] = \begin{Bmatrix} \exp(i\lambda T) \\ 0 \end{Bmatrix} ,\tau > T , \end{cases}\quad (4.3.3)$$

where

$$\chi = \sqrt{\lambda^2 + |q_0|^2} .$$

The expressions for the Jost coefficients are

$$a(\lambda) = e^{i\lambda T}\left(\cos\chi T - \frac{i\lambda}{\chi}\right)\sin\chi T ,$$

$$b(\lambda) = \frac{iq_0^*}{\chi}e^{-i\lambda T}\sin\chi T .\qquad (4.3.4)$$

Provided a coefficient $a(\lambda)$ is known, one can define the asymptotic behaviour of $|q(\xi,\tau)|^2$ for $\xi \gg 1$. Following *Zakharov* et al. (1980), we have

$$|q(\xi,\tau)|^2 = \frac{1}{4\pi\xi}\ln\left|a\left(-\frac{\tau}{4\xi}\right)\right|^{-2} .\qquad (4.3.5)$$

For the initial condition (4.3.2) with $q_0 = \text{const}$, $\varepsilon = 0$, we have

$$|a(\lambda)|^{-2} = \left(\lambda^2 + |q|^2\right)\left(\lambda_0^2 + |q_0|^2\cos^2\chi T\right)^{-1} .\qquad (4.3.6)$$

Let $\varepsilon(\tau)$ be a Gaussian random variable, with

$$\langle \varepsilon \rangle = 0 , \qquad \langle \varepsilon(\tau)\varepsilon(\tau') \rangle = \sigma_1^2 B_{\tau_1}(\tau - \tau') ,$$

$$\langle \varepsilon(\tau)\varepsilon^*(\tau') \rangle = \sigma_2^2 D_{\tau_2}(\tau - \tau') .$$

Here, τ_1 and τ_2 are correlation times. In the case of weak coherence to be treated below, one can apply perturbation theory. For this purpose, $\ln|a(\lambda)|^{-2}$ is first expanded in a Taylor series, and then averaged term-by-term over all the $\varepsilon(\tau)$. This gives

$$\langle |q|^2 \rangle = \frac{1}{4\pi\xi} \left(\left| \ln \left| a_0 \left(-\frac{\tau}{4\xi} \right) \right| \right|^2 + G(\lambda)|_{\lambda=4\nu} \right) ,$$

where

$$G(\lambda) = -\mathrm{Re}\left\{ \int_0^T d\tau \int_0^T d\tau' \left[\sigma_1^2 \left[\left(\frac{\psi^{(1)}(\tau)\psi^{(1)}(\tau)}{a^*} \right)^2 \right. \right. \right.$$

$$\left. \left. + \left(\frac{\psi^{*(2)}(\tau)\psi^{*(2)}(\tau')}{a^*} \right)^2 \right] B_{\tau_1}(\tau - \tau') \right.$$

$$\left. - \sigma_2^2 \left(\frac{\psi^{(1)}(\tau)\psi^{(2)}(\tau')}{a} \right)^2 + \left(\frac{\psi^{*(2)}(\tau)\psi^{*(1)}(\tau)}{a} \right)^2 \right] D_{\tau_2}(\tau - \tau') \right\} . \qquad (4.3.7)$$

The calculation shows that the correction for the initial pulse intensity at the maximum point $(\tau = 0)$ is

$$\frac{1}{2\pi} \sigma_2^2 \left(|q_0| \cos^2 |q_0 \bar{T}| \right)^{-1} \mathrm{Re}\left\{ \int_0^T d\tau F \left(|q_0|(\tau - T) \right) \right\} D_{\tau_2}(\tau) ,$$

$$F(x) = \sin 2x - \frac{x}{2} \left(1 + 2\cos^2 x \right) .$$

In the case of the Gaussian correlation function

$$D_{\tau_2}(\tau) = \exp\left(-\frac{\tau^2}{2\tau_2^2} \right)$$

for $\tau_2 \ll T, |q_0|^{-1}$, the above becomes

$$\frac{1}{\sqrt{2\pi}} \sigma_2^2 \tau_0 \left[q_0 \cos \left(|q_0^2|T \right) \right]^{-1} \left[\sqrt{2}\cos \left(2|q_0|T + \frac{\pi}{4} \right) + |q_0|T \right] .$$

The mean pulse intensity is

$$\langle I \rangle = \int_0^T d\tau \langle |q(\tau, 0)|^2 \rangle = |q_0|^2 T + \sigma_0^2 D(0)T .$$

It means that the condition for soliton generation is

$$\langle I \rangle < \sigma_2^2 D(0)T + I_{cr} ,$$

i.e., the threshold is increased.

Let us consider in more detail the problem of soliton generation from a random initial potential (*Bass* et al., 1988). Let the initial potential have the form

$$q(\xi = 0, \tau) = \begin{cases} q_0(\tau) & 0 < \tau < \tau_0 \\ 0 & \tau < 0, \tau > \tau_0 \end{cases},$$

where $q_0(\tau)$ is a Gaussian random process, satisfying

$$\langle q_0 \rangle = p, \quad \langle q_0(\tau)q_0(\tau') \rangle = B(\tau - \tau'). \tag{4.3.8}$$

The threshold condition for soliton generation is

$$\theta(\tau) = \int_0^\tau |q_0(\tau')| d\tau' \geq \frac{\pi}{2}.$$

The probability for generating a soliton is therefore equal to the probability that $\theta(\tau) > \pi/2$. Hence, the probability for soliton generation is

$$P\left(\theta > \frac{\pi}{2}\right) = 1 - \mathrm{erf}\left[\frac{\pi/2 - p\tau}{\sigma\theta}\right],$$

where $\mathrm{erf}(x)$ is the error function. We take the correlator in the form

$$B(t - t') = \sigma_q^2 t_c^{-1} \exp\left(\frac{|t - t'|}{t_c}\right),$$

where t_c is the correlation time of $q_0(\tau)$. Then for $\sigma_\theta^2(\sigma_q, t_c)$ the expression

$$\sigma_\theta^2(\sigma_q, t_c) = \sigma_q^2 \tau \left(1 - \frac{1 - \exp(-\tau/t_c)}{\tau/t_c}\right)$$

is obtained. This expression shows that the probability of soliton generation grows for $\tau_0/t_c \to \infty$ and decreases for $\tau_0/t_c \to 0$. Such a result is evident because the input intensity is inversely proportional to t_c^{-1} and the probability of soliton generation is proportional to the pulse amplitude.

4.4 Random Pulses in Fibres (Soliton Region)

The investigation of short pulse propagation along optical fibres requires that the incoherence of input pulses must be taken into account. The corresponding random parameter modulation can naturally be divided into two types: amplitude and phase modulation.

Let us first analyze the results of numerical simulations of the propagation of incoherent pulses along fibres (*Maimistov* et al., 1986; *Manykin* et al., 1986; *Klovsky et al*, 1987).

The evolution of two types of initial random pulses has been considered. The first type dealt with the amplitude noise, when

$$q_0(\tau) = \bar{q}_0(\tau) + \alpha(\tau) .$$

The second one corresponded to a case of the phase noise, i.e.,

$$q_0(\tau) = \bar{q}_0(\tau) \exp[(i\varphi)] .$$

Here, the functions $\alpha(\tau)$ and $\varphi(\tau)$ were chosen in the form of the δ-correlated Gaussian random functions

$$\langle \alpha(\tau) \rangle = 0 , \quad \langle \alpha(\tau)\alpha(\tau') \rangle = \sigma_a^2 \delta(\tau - \tau') ,$$

$$\langle \varphi(\tau) \rangle = 0 , \quad \langle \varphi(\tau)\varphi(\tau') \rangle = \sigma_\varphi^2 \delta(\tau - \tau') .$$

The numerical simulation of an evolution $\bar{q}_0(\tau) = \mathrm{sech}\tau$ in the presence of the amplitude and phase noise at different values σ_a^2 and σ_φ^2 has been performed.

When $\sigma_a^2 = 0.05$, $\sigma_\varphi^2 = 0$, a pulse propagated up to a distance of $\xi = 30$ without broadening and, at the same time, a weak random amplitude modulation was preserved. This behaviour was observed up to $\sigma_a^2 \leq 0.5$. When $\sigma_a^2 > 0.5$, the modulation amplitude increased, the pulse broadened, and its decay took place. When the initial pulse was small, e.g., $\bar{q}_0(\tau) = 0.5 \,\mathrm{sech}\tau$, the influence of $\alpha(\tau)$ resulted in very rapid signal decay. So, while propagating over a distance of about $5\ell_d$, the pulse amplitude decreased by three times.

The numerical simulation showed also that the pulse evolution under the action of phase fluctuations was slightly different from the influence of amplitude fluctuations. This phenomenon reflects the nonlinear character of the equation for $q(\xi, \tau)$. The phase modulation changes already for small distances into amplitude modulation. The analysis has demonstrated that pulse broadening and growth of amplitude modulation are observed when $\sigma_\varphi^2 = 0.5$. Further, it is also essential that, for different realizations of the random phase, modulation results in different values of the velocity of the generated solitons. The occurrence of fluctuations in the soliton velocity when the initial condition has a complex Gaussian noise term was also discovered by *Manykin* et al. (1987). They considered two types of initial conditions:

$$q_0(\tau) = \sigma\varepsilon(\tau) \, e^{-\tau^2/2} ; \quad q_0(\tau) = e^{-\tau^2/2} + \sigma\varepsilon(\tau) ,$$

where $\varepsilon(\tau)$ is a complex Gaussian noise term:

$$\langle \varepsilon \rangle = 0 , \quad \langle \varepsilon(\tau)\varepsilon^*(\tau + \theta) \rangle = \exp\left(-\frac{\theta^2}{\tau_c^2}\right) .$$

For $\sigma = 0.1 \sim 0.4$; $\tau = 0.4$, velocity fluctuations were observed which grow with σ. *Mork* et al. (1987) studied the evolution of pulses containing a fluctuating phase $\theta(t)$ characteristic of those generated by semiconductor or glass lasers. The model for $\theta(t)$ takes the form

$$\frac{d\theta}{dt} = F(t) , \quad \langle F(t) \rangle = 0 .$$

Here, $F(t)$ is a random force, which is δ-correlated:

$$\langle F(t_1)F(t_2) \rangle = 2D\delta(t_1 - t_2) .$$

The value $D = 1.10^{-4}$ corresponds to the laser line width $D/\pi\tau = 32$ MHz for $\tau_0 = 1$ ps. Using the numerical simulation, researchers have investigated a distribution of eigenvalues obtained from the statistical ensemble of pulses. The Gaussian distribution has a zero mean and standard deviation $\sigma = \sqrt{2D\Delta t}$, where Δt is the grid spacing. The averaging has been performed over 1000 pulses. Numerical simulation showed that stochastic phase modulation leads to a small increase of an averaged imaginary part of the spectral parameter λ and the soliton amplitude η, respectively.

Hence, it follows that the generated solitons are stable even with stochastic variations of pulse phase caused by the finiteness of the laser width.

The distribution of the real parts of the eigenvalues corresponds to a distribution of the soliton velocity (-2ζ) and gives rise to a jitter $\Delta t'$ in the propagation time of the pulses. Simulation showed a jitter of $\Delta t' = 2$ ps for $z = 100$ km and the standard deviation of the velocity distribution $\Delta\zeta = 4.2 \cdot 10^{-3}$.

The case when the initial pulse has the form of "soliton + noise" has been considered by *Vysloukh* and *Ivanov* (1988). The initial pulse has the form

$$q_0(\tau) = q_0^*(\tau) \left[1 + \bar{\xi}(\tau) \right] ,$$

where $\bar{\xi}(\tau)$ is a Gaussian random process,

$$\bar{\xi} = \xi_1(\tau) + i\xi_2(\tau) , \quad \langle \bar{\xi} \rangle = 0 ,$$

$$\langle \bar{\xi}(\tau)\bar{\xi}(0) \rangle = 2\sigma^2 \exp\left(-\frac{\tau^2}{\tau_c^2} \right) ,$$

and τ_c is the correlation time. The dependence of subsequent evolution on the noise amplitude and correlation time has been studied. It was found that for $q_0 = 3$ sechτ and $\sigma > 0.2$ an irreversible decay of the soliton bound state occurred.

We will now proceed with an analytical consideration of the latter problem (*Abdullaev* and *Darmanyan*, 1989). Consider the NLS equation

$$iq_\xi + q_{\tau\tau} + 2|q|^2 q = 0 ,$$

with the initial condition

$$q(\tau, \xi = 0) = q_s(\tau)(1 + \varepsilon_1(\tau))e^{i\varepsilon_2(\tau)} , \tag{4.4.1}$$

where

$$q_s(\tau) = 2\eta \, \text{sech} \, 2\eta(\tau - \tau_0) \exp[-2i\mu\tau - i\theta_0] \tag{4.4.2}$$

and $\varepsilon_{1,2}$ are the Gaussian random functions,

$$\langle \varepsilon_{1,2} \rangle = 0 \;,$$

$$\langle \varepsilon_{1,2}(\tau) \varepsilon_{1,2}(\tau') \rangle = B_{1,2}(\tau - \tau'; \tau_{1,2}) \;. \tag{4.4.3}$$

We will use the IST to study the evolution of $q(\tau, \xi)$ subject to the initial condition (4.4.1). According to the IST scheme, the Cauchy problem is reduced to an investigation of the spectral features of the linear problem (4.3.1).

When $\varepsilon_{1,2} \neq 0$ in (4.4.1), a reflected wave is scattered from the potential q, corresponding to the generation of continuous "noise". To proceed further, we need the Jost functions; these have the form

$$\phi(\tau, \lambda) = \frac{e^{i\lambda\tau}}{\lambda - \mu + i\eta} \begin{pmatrix} \lambda - \mu + i\eta \tanh 2\eta(\tau - \tau_0) \\ \eta \operatorname{sech} 2\eta(\tau - \tau_0) \end{pmatrix} \;, \tag{4.4.4}$$

$$\phi(\tau, \lambda) = \frac{e^{-i\lambda\tau}}{\lambda - \mu + i\eta} \begin{pmatrix} -\eta \operatorname{sech} 2\eta(\tau - \tau_0)e^{2i\mu\tau + i\theta_0} \\ \lambda - \mu - i\eta \tanh 2\eta(\tau - \tau_0) \end{pmatrix} \;. \tag{4.4.5}$$

To define parameters of the continuous spectrum the coefficient $b(\lambda)$ must first of all be evaluated. The variational derivatives (*Zakharov* et al., 1980) are

$$\frac{\delta b(\lambda)}{\delta q(x)} = -i\varphi^{(1)} \tilde{\phi}^{(2)} \;, \qquad \frac{\delta b(\lambda)}{\delta q^*(x)} = i\varphi^{(2)*} \tilde{\phi}^{(2)} \;, \tag{4.4.6}$$

where

$$\tilde{\phi} = \begin{pmatrix} \tilde{\phi}^{(1)} \\ \tilde{\phi}^{(2)} \end{pmatrix} = \begin{pmatrix} -\phi^{(2)*} \\ \phi^{(1)*} \end{pmatrix}$$

will be required. From this, we derive

$$b(\lambda) = \frac{i\exp(-2i\mu\eta\tau_0 + i\theta_0)}{1 + \kappa^2}$$
$$\times \int_{-\infty}^{\infty} \frac{dz}{\cosh z} \exp\left[f^*(\mu - i\tanh z)^2 - f\operatorname{sech}^2 z \right] \;, \tag{4.4.7}$$

where

$$\kappa = \frac{\lambda - \mu}{\eta} \;, \qquad z = 2\eta(\tau - \tau_0) \;,$$

$$f(\tau) = (1 + \varepsilon_1(\tau))e^{i\varepsilon_2(\tau)} - 1 \;.$$

Now an expression for $\langle |b(\lambda)|^2 \rangle$ is easily obtained:

$$\langle |b(\lambda)|^2 \rangle \approx 4\sqrt{\pi}\eta\tau_1 B_{01} \left[1 - \frac{4}{15(1 + \kappa^2)} \right] \exp\left(-\frac{\kappa\tau_1^2}{4} \right)$$
$$+ 4\sqrt{\pi}\eta\tau_2 B_{02} \left[1 - \frac{8}{3(1 + \kappa^2)} + \frac{5}{(1 + \kappa^2)} \right]$$
$$\times \exp\left(-\frac{\kappa^2\tau_2^2}{4} \right) \;. \tag{4.4.8}$$

We have assumed that

$$B_i(\tau - \tau'; \tau_i) = B_{0i} \exp\left[-\frac{(\tau - \tau')^2}{\tau_i^2}\right] , \quad \tau_i \ll 1 .$$

Using this expression, we can evaluate the contribution of the continuous spectrum to the NLS integral invariants. For the energy (for the case $|b(\lambda)| \ll 1$), this gives

$$I_1 = I_1^d + I_1^c = \int_{-\infty}^{\infty} |q|^2 d\tau = 4\text{Im}\{\lambda\} + \frac{1}{\pi} \int_{-\infty}^{\infty} d\lambda \, n(\lambda) . \tag{4.4.9}$$

The contribution from the continuum spectrum is given by

$$\langle I_1^c \rangle \approx \frac{1}{\pi} \int_{-\infty}^{\infty} d\lambda \langle |b|^2 \rangle$$
$$\approx 4\eta \left[B_{01}(1 - 0.12\eta\tau_1 - 2.4\eta^2\tau_1^2) \right.$$
$$\left. + B_{02}(1 - 2.6\eta\tau_2 + 3.6\eta^2\tau_2^2) \right] . \tag{4.4.10}$$

Taking into account $\langle I_1^d \rangle \approx 4\eta$ (see (B.11)) and $2\eta\tau_2 \approx \tau_2/\tau_s \ll 1$, we obtain the ratio of the energy of the radiation field to the one of the soliton

$$\frac{\langle I_1^c \rangle}{\langle I_1^d \rangle} \approx B_{01} + B_{02} .$$

We now find the change in parameters of the discrete spectrum $\Delta\lambda = \Delta\mu + i\Delta\eta$ which characterize the change in soliton velocity and amplitude. In order to do so, we require formulae for the variational derivatives of the spectral parameter λ reported in Appendix B (B.33–38), which are to be substituted by

$$\lambda = \lambda_0 + \int_{-\infty}^{\infty} d\tau \left(\delta q \frac{\delta\lambda}{\delta q} + \delta q^* \frac{\delta\lambda}{\delta q^*} \right)$$
$$+ \frac{1}{2} \iint_{-\infty}^{\infty} d\tau_1 d\tau_2 \left[\delta q(\tau_1)\delta q(\tau_2) \frac{\delta^2\lambda}{\delta q(\tau_1)\delta q(\tau_2)} \right.$$
$$\left. + 2\delta q(\tau_1)\delta q^*(\tau_2) \frac{\delta^2\lambda}{\delta q(\tau_1)\delta q^*(\tau_2)} \right] . \tag{4.4.11}$$

Taking into account that $\delta q = q_s f(\tau)$, where $f(\tau)$ is defined above to the second order, we use

$$f(\tau) \approx \varepsilon_1 - \frac{\varepsilon_2^2}{2} + i(\varepsilon_2 + \varepsilon_1\varepsilon_2)$$

to obtain

$$\Delta\mu = \eta \int_{-\infty}^{\infty} dz \, \tanh z \, \text{sech}^2 z \, \text{Im}\{f(z)\} ,$$

$$\Delta \eta = \eta \int dz \, \text{sech} z \, \text{Re} \, \{f(z)\}$$

$$+ \int \int dz_1 dz_2 \frac{\sinh z_2 - \sinh z_1}{\cosh^3 z_1 \cosh^3 z_2} \, \varepsilon(z_1)\varepsilon(z_2)\theta(z_1 - z_2)$$

$$+ \frac{1}{2} \int \int \frac{dz_1 dz_2}{\cosh^2 z_1 \cosh^2 z_2} \left[\varepsilon_1(z_1)\varepsilon(z_2) \right.$$

$$\left. + \tanh z_1 \tanh z_2 \varepsilon(z_1)\varepsilon(z_2) \right] . \tag{4.4.12}$$

When $\tau_i \ll 1$, we find

$$\langle \Delta \mu \rangle = 0 \,,$$

$$\langle \Delta \mu^2 \rangle \approx \frac{4\sqrt{\pi}}{15} \eta^3 \tau_2 B_{02} \,, \tag{4.4.13}$$

$$\langle \Delta \eta \rangle \approx \eta B_{02} \left[-1 + \frac{4\sqrt{\pi}}{15} \eta \tau_2 - 3\eta^2 \tau_2^2 \right] + \frac{4\sqrt{\pi}}{3\eta^2 \tau_1} B_{01} \,,$$

$$\langle \Delta \eta^2 \rangle \approx \frac{4\sqrt{\pi}}{3} \eta^3 \tau_1 B_{01} \,.$$

As seen from (4.4.12), the second-order correction to λ is purely imaginary, i.e., it gives no contribution to the velocity change. This is in agreement with results obtained by *Lewiz* et al. (1986) and *Elgin* (1985), who used a random phase modulation of the initial pulse. If the random function $\varepsilon(\tau)$ is Gaussian, it follows that the distribution function for the spectral parameter will also be Gaussian. An analogous investigation can be carried out for the evolution of a randomly modulated dark soliton, using the analogy between the KdV and the NLS equations accounting for positive group dispersion (*Gredeskul* et al., 1990; *Abdullaev* and *Darmanyan*, 1988; *Kivshar*, 1991). The randomly modulated multi-soliton initial state was considered by *Konotop* (1991).

In nonlinear optics, the sine-Gordon equation

$$u_{tt} - u_{xx} + \sin u = 0 \tag{4.4.14}$$

arises, which has the one-soliton solution:

$$u_s(x, t) = 4 \tan^{-1}[\exp(\sigma z)] \,,$$

$$z = \frac{x - x_0 - vt}{\sqrt{1 - v^2}} \,. \tag{4.4.15}$$

Here, $\sigma = \pm 1$ corresponds to a soliton (kink) or an antisoliton (antikink), respectively. Using techniques just discussed, we can consider the evolution of a randomly modulated initial potential of the type

$$u_t(x, t = 0) = u_{st}(x, t = 0)[1 + \varepsilon(x)] \,,$$

where ε is a Gaussian random function with a correlation length ℓ_c (*Abdullaev* and *Darmanyan*, 1989). It is found that, for $\ell_c v \ll 1$, the spectral density of the energy is maximal for values of wave number $k \sim 1$, and that the mean value of the emitted energy is equal to $\langle H^c \rangle \approx \pi B_0 v^2$.

4.5 Solitons in Random Nonuniform Fibres

In this section, we discuss the effects of nonuniformities, which are always present in real fibres, on soliton dynamics. Consider the case when nonuniformities are distributed along the direction of pulse propagation in the fibre. Two types of nonuniformity are considered: (i) that arising from a variation in parameters of the fibre medium, and (ii) from a variation of fibre geometry (diameter fluctuations, etc.).

An equation describing optical pulse propagation through a nonuniform fibre has, in dimensionless variables, the following form:

$$iq_\xi + \frac{1}{2}\, q_{\tau\tau} + [1 + \alpha_1(\xi)]\,|q|^2 q + \alpha_2(\xi)q = 0 \ . \tag{4.5.1}$$

Here, $\alpha_1, \alpha_2 \ll 1$ describe fluctuations in the diameter, and in the dielectric properties of the fibre core, respectively (*Abdullaev*, 1983). Fibres, used in practice, typically have $\langle |\alpha_1|^2 \rangle \sim 10^{-1}$. The function $\alpha_2(\xi)$ is related to index fluctuations along the fibre due to nonuniformities in the material by the formula

$$\alpha_2 = \frac{2\Delta\varepsilon}{\varepsilon_2 |E_{\max}|^2} \ , \tag{4.5.2}$$

where $\Delta\varepsilon$ is the fluctuating part of the dielectric permeability ($\varepsilon = \varepsilon_0 + \Delta\varepsilon + \varepsilon_2 E^2$) of the fibre.

In the following, the random functions α_1 and α_2 will be assumed to be Gaussian distributed with zero mean values

$$\langle \alpha_1 \rangle = \langle \alpha_2 \rangle = 0 \ , \tag{4.5.3}$$

and with correlation functions

$$B_{1,2}(\xi, \xi') = \langle \alpha_{1,2}(\xi)\alpha_{1,2}(\xi') \rangle = \sigma_{1,2}e^{-D_{1,2}|\xi-\xi'|} \ . \tag{4.5.4}$$

We now examine (4.5.1) using the perturbation theory developed for solitons (*Abdullaev* et al., 1986; *Abrarov* and *Darmanyan*, 1986). Within the adiabatic approximation we obtain the following equations for the soliton parameters

$$q_s = q_0 \mathrm{sech}\,[q_0(\tau - \eta)] \exp\left\{ i \left(\frac{q_0^2 \xi}{2} + \delta \right) \right\} \ , \tag{4.5.5}$$

$$\frac{dq_0}{d\xi} = 0 \ , \quad \frac{d\eta}{d\xi} = 0 \ , \tag{4.5.6}$$

$$\frac{d\delta}{d\xi} = q_0^2 \alpha_1 + \alpha_2 \ . \tag{4.5.7}$$

As inferred from (4.5.7), a nonuniform fibre behaves as a phase screen, modulating only the phase and leaving the wave amplitude unchanged. The transmission function is obtained as

$$f(\xi) = \exp\left\{i\frac{q_0^2\xi}{2} + i\,q_0^2\int_0^\xi \alpha_1(\xi)d\xi + i\int_0^\xi \alpha_2(\xi)d\xi\right\}\,.$$

With the notation

$$I_1(\xi) = \int_0^\xi \alpha_1(\xi)d\xi\,, \quad I_2(\xi) = \int_0^\xi \alpha_2(\xi)d\xi\,,$$

the mean field is evaluated to be

$$\langle q \rangle = q_0 \mathrm{sech}\,[q_0(\tau - \eta)]\exp\left[i\frac{q_0^2\xi}{2}\right]\langle e^{i(q_0^2 I_1 + I_2)}\rangle\,. \tag{4.5.8}$$

Functions I_1 and I_2 are linearly related to the Gaussian random functions α_1 and α_2, and are therefore also Gaussian. Hence, the expression

$$\langle e^{i(q_0^2 I_1 + I_2)}\rangle = e^{(-1/2)\langle(q_0^2 I_1 + I_2)^2\rangle} = e^{(-1/2)(q_0^4\langle I_1^2\rangle + \langle I_2^2\rangle)} \tag{4.5.9}$$

is valid (*Rytov*, 1976), where I_1 and I_2 have been treated as statistically independent processes. The magnitudes $\langle I_1^2 \rangle$ and $\langle I_2^2 \rangle$ can be evaluated in terms of the correlation functions $B_{1,2}(\xi, \xi')$:

$$\langle I_{1,2}^2\rangle = \int_0^\xi d\xi'\int_0^\xi d\xi''B_{1,2}(\xi', \xi'') = 2\int_0^\xi (\xi - \chi)B_{1,2}(\chi)d\chi\,. \tag{4.5.10}$$

Equations (4.5.4,10) then show that

$$\langle I_{1,2}^2\rangle = \frac{2\sigma_{1,2}^2}{D_{1,2}^2}\left(D_{1,2}\xi - 1 + e^{-D_{1,2}\xi}\right)\,. \tag{4.5.11}$$

Using (4.5.8,9) for the mean field, we obtain

$$\langle q \rangle = q_0 \mathrm{sech}\,[q_0(\tau - \eta)]\exp\left[i\frac{q_0^2\xi}{2}\right]$$
$$\times \exp\left\{-\frac{q_0^4\sigma_1^2}{D_1^2}\left(D_1\xi - 1 + e^{-D_1\xi}\right) - \frac{\sigma_2^2}{D_2^2}\left(D_2\xi - 1 + e^{-D_2\xi}\right)\right\}\,. \tag{4.5.12}$$

In the limiting case when $D_{1,2}\xi \gg 1$, (4.5.12) becomes

$$\langle q \rangle = q_0 \mathrm{sech}\,[q_0(\tau - \eta)]\exp\left[i\frac{q_0^2\xi}{2}\right]\exp\left\{-\left(\frac{q_0^4\sigma_1^2}{D_1} + \frac{\sigma_2^2}{D_2}\right)\xi\right\}\,. \tag{4.5.13}$$

As seen from this expression, the mean field decreases exponentially. A similar result can be obtained using the mean field method based on the solution of moment equations (*Klyazkin*, 1980; *Ishimaru*, 1978).

Note that this decrease is associated with ensemble-averaging, whereas in each separate realization a soliton experiences only the phase shift (4.5.7).

The mean values for intensity and energy density are expressed by correlation functions of the second order. For $\xi' < \xi$:

$$\Gamma_1(\xi, \xi', \tau) = \langle q(\xi, \tau) q(\xi', \tau) \rangle$$

$$= q_0^2 \text{sech}^2 \left[q_0(\tau - \xi) \right] \exp \left[i q_0^2 \frac{\xi + \xi'}{2} \right]$$

$$\times \exp \left\{ -\frac{q_0^4 \sigma_1^2}{D_1^2} \left[2 D_1 (2\xi + \xi') - 5 + 3 \left(e^{-D_1 \xi} + e^{-D_1 \xi'} \right) \right.\right.$$

$$\left. - e^{-D_1(\xi' - \xi)} \right] - \frac{\sigma_2^2}{D_2^2} \left[2 D_2 (2\xi + \xi') - 5 \right.$$

$$\left.\left. + 3 \left(e^{-D_2 \xi} + e^{-D_2 \xi'} \right) - e^{-D_2(\xi' - \xi)} \right] \right\} . \tag{4.5.14}$$

For $\xi' < \xi$:

$$\Gamma_2(\xi, \xi', \tau) = \langle q(\xi, \tau) q^*(\xi', \tau) \rangle = q_0^2 \text{sech}^2 \left[q_0(\tau - \xi) \right]$$

$$\times \exp \left\{ -\frac{q_0^4 \sigma_1^2}{D_1^2} \left[D_1 |\xi - \xi'| - 1 + e^{-D_1 |\xi' - \xi|} \right] \right.$$

$$\left. - \frac{\sigma_2^2}{D_2^2} \left[D_2 |\xi - \xi'| - 1 + e^{-D_2 |\xi' - \xi|} \right] \right\} .$$

The quantities $\Gamma_{1,2}$ are in accordance with an experimental definition of fibre nonuniformity.

It is also possible to consider a more general case when the dielectric constant is modulated in ξ and τ,

$$\varepsilon = \varepsilon_0 + \varepsilon_1(\xi, \tau) .$$

Such models are associated with changes in ε induced by sound waves. The propagation of short pulses is then described by the stochastic NLS equation

$$i q_\xi + q_{\tau\tau} + |q|^2 q = \tilde{\varepsilon}(\xi, \tau) q - i \gamma q ,$$

where ε corresponds to a Gaussian random process. *Bass* et al. (1989) investigated the dynamics of optical solitons within the framework of an adiabatic perturbation theory, and derived an appropriate probability density $P(\mu, \zeta; \xi)$. The calculation of the mean soliton intensity $\langle |q|^2 \rangle$ gives

$$\langle |q|^2 \rangle = 8 \eta_0^2 \sum_{n=1}^{\infty} (-1)^n n \exp(na + n^2 b) ,$$

where $a = 4\eta_0(\tau - \mu_0 \xi)$, $b = (64/3) D \xi^3 \eta_0$, and $D = \eta_0^2$ is the diffusion coefficient.

It is clear that the mean soliton intensity transforms into a Gaussian wave packet during propagation in the fibre (see also *Abdullaev*, 1982).

Consider now the propagation of a SG soliton and a breather over a medium with random nonuniformities (*Umarov*, 1986; *Bass* et al., 1986). This problem is of importance, not only in the study of optical solitons in a nonuniform resonance medium, but also from a more general viewpoint since it combines investigations of the dynamics of domain walls in, for example, ferromagnetics and vortices in long Josephson junctions (*Abdullaev and Khabibullaev*, 1986).

A wave equation describing the motion of a soliton in the field of a random potential has the form

$$u_{tt} - u_{xx} + \sin u = \varepsilon(x) R(u) , \tag{4.5.15}$$

where u is a normalized field strength, $\varepsilon(x) R(u)$ describes the random nonuniformity of a medium and it is assumed that $\varepsilon << 1$. We will consider $\varepsilon(x)$ as a Gaussian random function with an exponential correlator

$$\langle \varepsilon \rangle = 0 , \quad \langle \varepsilon(x) \varepsilon(x') \rangle = \frac{\alpha}{2 \ell_\varepsilon} \exp\left[-\frac{|x - x'|}{\ell_\varepsilon}\right] , \tag{4.5.16}$$

where α is the power of the random potential and ℓ_ε the correlation lenght.

A soliton solution of (4.5.15) for $\varepsilon = 0$ is given by (4.4.15).

The motion of the soliton driven by random potential resembles the motion of a classical particle subjected to random forces. The relevant adiabatic dynamics have been studied in detail by *Abdullaev* et al. (1983). Here, we study another aspect of the problem associated with the radiation emitted by a soliton moving in a random inhomogeneous medium (*Umarov*, 1986; *Kivshar* et al., 1986).

The required radiation power can be calculated using the formula (*Zakharov* et al., 1980)

$$E = \frac{4}{\pi} \int_0^\infty d\lambda \left(1 + \frac{1}{4} \lambda^{-2}\right) |b(\lambda, t)|^2 , \tag{4.5.17}$$

where $b(\lambda, t)$ can be derived from (*Kaup* and *Newell*, 1978)

$$\frac{db}{dt} = -2i\omega(\lambda)b + \frac{i}{4} \int_0^\infty R(u_0)\varepsilon(x)\left[\varphi_1(x, t)\phi_2^*(x, t)\right.$$
$$\left. + \varphi_2(x, t)\phi_1^*(x, t)\right] dx . \tag{4.5.18}$$

Here, $\varepsilon R(u_0)$ describes the perturbation and ϕ, φ are the Jost functions of the unperturbed SG equation. From (4.5.17) it follows that

$$\frac{dE}{dt} = \frac{16}{\pi} \int_0^\infty d\lambda \frac{\omega(\lambda)}{\lambda} \mathrm{Re}\left\{b^* \frac{db}{dt}\right\} , \tag{4.5.19}$$

where

$$\omega(\lambda) = \frac{1}{2}\left(\lambda + \frac{1}{4\lambda}\right) .$$

Solving (4.5.19) with the aid of (4.5.18), one can obtain an expression for the soliton radiation power (for the case $R(u) = u_{xx}$)

$$\frac{d\langle E\rangle}{dt} = \frac{2\alpha^2}{\pi^3\gamma^2} \int_1^\infty \frac{d\omega\, f(\omega)}{k\cosh^2(\pi\omega\gamma/2v_1)}\,, \qquad (4.5.20)$$

where

$$f(\omega) = \frac{\left\{\left[v_1^2 - (\omega + k)^2/4\right]\gamma\omega/v - v_1(\omega + k)(1 - \omega^2\gamma^2/v^2)\right\}^2}{(\lambda^2 + v_1^2)(1 + \kappa_0^2\ell_\varepsilon^2)}\,,$$

$$\kappa_0^2 = k + \frac{\omega}{v}\,, \quad v_1 = \frac{1}{2}\sqrt{\frac{1 + v}{1 - v}}\,,$$

$$k = \sqrt{\omega^2 - 1}\,, \quad \gamma = \sqrt{1 - v^2}\,.$$

As seen from (4.5.20), the radiation spectrum contains all frequencies beginning with the threshold $\omega = 1$. As $v \to 0$, the integral in (4.5.20) can be evaluated to give

$$\frac{d\langle E\rangle}{dt} = \frac{\alpha^2 e^{-\pi/v}}{\sqrt{2}v^{7/2}\pi^{5/2}(1 + \ell_\varepsilon^2/v^2)}\,. \qquad (4.5.21)$$

This implies that the soliton radiation power in a randomly-inhomogeneous medium is exponentially small as $v \to 0$. For $v \to 1$, we obtain

$$\frac{d\langle E\rangle}{dt} \sim \frac{2\alpha^2}{\pi^3(1 - v^2)}\,, \qquad (4.5.22)$$

i.e., the radiation power increases. This behaviour was observed in numerical simulations of the stochastically perturbed SG equation by *Khikmatov* (1987).

Very recently essential new results have been obtained for the stability of soliton propagation in fibers with random birefringence (*Wai* et al., 1991) and soliton instabilities from resonant random mode coupling in birefringent fibers (*de Angelis* et al., 1992).

4.6 Random Amplification of Solitons

We now consider the propagation of an optical soliton along an active fribre with random nonuniformities. Such problems are relevant to the study of stable information transmission over long distances ($\geq 10^3$ km) using solitons. Several experimental schemes propose to install uniformly spaced amplifiers in a long optic-fibre cable to compensate for fibre loss. In real conditions, the amplification value will be different and random for the different amplifiers. In this section we will study the effect of gain fluctuations along the fibre on the stability of optical solitons and determine the modes of their effective propagation over long distances.

Consider the propagation of an optical soliton over a fibre with periodically spaced amplifiers (*Kodama* and *Hasegawa*, 1983). The relevant wave equation is the perturbed NLS equation

$$iq_\xi + \frac{1}{2} q_{\tau\tau} + |q|^2 q = -i\Gamma q + i \sum_{\ell=1}^N \alpha q(\tau, \xi_\ell - 0)\delta(\xi - \xi_\ell) . \qquad (4.6.1)$$

Here, $\Delta\xi$ is the spacing between the amplifiers, $\xi_\ell = (\ell - 1)\Delta\xi$, $\ell = 1, 2, \ldots, N$ labels the amplifiers, and α_ℓ is the gain coefficient of the ℓ-th amplifier. The coefficients α_ℓ are assumed to be random.

We now examine the variation of the mean energy during the soliton propagation in the fibre. It follows from (4.6.1) that the soliton energy is

$$E(\xi) = \int_{-\infty}^{\infty} |q(\xi, \tau)|^2 d\tau = F^2(\xi)E_0 , \qquad (4.6.2)$$

where

$$\xi_n < \xi < \xi_{n+1} ,$$

$$F(\xi) = \left[\prod_{\ell}^{n}(1 + \alpha_\ell) \right] \exp(-\Gamma\xi) .$$

Function $F(\xi)$ satisfies

$$\frac{\partial F}{\partial \xi} = -\Gamma F + \sum_{1}^{N} \alpha_\ell F(\xi_\ell - 0)\delta(\xi - \xi_\ell) . \qquad (4.6.3)$$

It is clear that $F(\xi)$ is a Markov random process

$$F_n = F_n(\xi_n + 0) = (1 + \alpha_n)e^{-\Gamma\Delta\xi} F(\xi_{n-1} - 0) = (1 + \beta_n)F_{n-1}$$
$$F_0 = e^{\Gamma\Delta\xi} , \qquad (4.6.4)$$

where

$$\beta_n = (\alpha_n - \bar{\alpha})e^{-\Gamma\Delta\xi} , \qquad \bar{\alpha} = e^{\Gamma\Delta\xi} - 1 ;$$

$\bar{\alpha}$ is the value for gain which exactly compensates loss.

Using a standard technique, we obtain the Fokker-Planck equation for the probability density $P(F, \xi)$:

$$\frac{\partial P}{\partial \xi} = D \left(\frac{\partial^2}{\partial F^2} \right) (F^2 P) ,$$

where

$$D = \lim_{\Delta\xi \to 0} \frac{\langle (\alpha_n - \bar{\alpha})^2 \rangle}{2\Delta\xi} \qquad (4.6.5)$$

is the diffusion constant. Given an initial condition of the form

$$P(F, 0) = \delta(F - F_0) ,$$

the solution of (4.6.5) is

$$P(F, \xi) = \frac{1}{\sqrt{4\pi\xi D}\,F} \exp\left\{ -\frac{1}{4D\xi} \left(\log\frac{F}{F_0} + D\xi \right)^2 \right\} . \qquad (4.6.6)$$

It follows that the total mean energy is given by

$$\langle E(\xi) - E(0) \rangle = \left(e^{2D\xi} - 1 \right) E(0) . \qquad (4.6.7)$$

We now report simulation results on the NLS soliton dynamics under the action of random amplification. As an example, we consider a 34.2-ps soliton with peak power of 11.2 W propagating over a 0.2-dB/km-loss fibre with a spacing between the amplifiers of 10 km. To maintain a stationary mode the gain should be $\bar{\alpha} = 0.21$. A calculation was carried out for two values of $\sqrt{2D}$; namely, 0.025 and 0.05 (see Fig. 4.5).

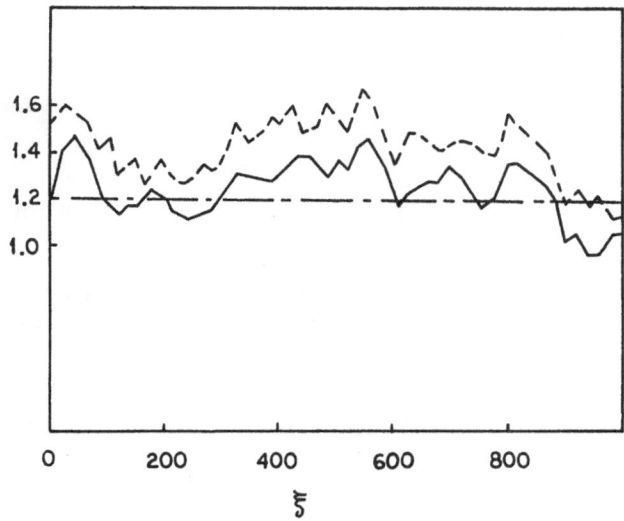

Fig. 4.5. A pulse amplitude and energy variation is shown by solid and dotted lines, respectively. The fibre loss is 0.2 dB/km, $\bar{\alpha} = 0.21$, $\sqrt{2D} = 0.025$, the amplifier spacing is 10 km, $2t_0 = 34.2$ ps, $P_0 = 11.2$ mW. From *Kodama* and *Hasegawa* (1983)

Simulations showed that the soliton was stable, and that changes of its profile appeared only as weak random modulations of its tail (*Kodama* and *Hasegawa*, 1983). We conclude that the influence of fluctuation in the gain parameter is small, and that the soliton will propagate in fibres for thousands of kilometers with small losses.

Let us analyze the amplification of solitons by a parametrically acting random field (*Abdullaev* et al., 1986; 1989). Such problems arise in the study of the interaction between an SRS soliton in a single-mode fibre with an incoherent pump field (*Dianov* et al., 1985). The evolution equation is

$$iq_\xi + \frac{1}{2} q_{\tau\tau} + |q|^2 q = -i\Gamma q + i\varepsilon(\tau, \xi)e^{-\Gamma_1 \xi} q \ . \tag{4.6.8}$$

Using adiabatic perturbation theory the following equation is obtained for the soliton amplitude:

$$\eta_\xi = \eta \left[-2\Gamma + \exp(-\Gamma_1 \xi) \int_{-\infty}^{\infty} \varepsilon \left(\frac{z}{2\eta} + \tau_0 \xi \right) \text{sech}^2 z \, dz \right] \ .$$

Integrating, and then taking an ensemble average gives the following expression for the RMS soliton amplitude:

$$\langle \eta^2 \rangle = \eta_0^2 \exp(-4\Gamma\xi)$$
$$\times \left\langle \exp \left[2 \int_{-\infty}^{\infty} d\xi_1 \exp(-\Gamma_1 \xi) \int_{-\infty}^{\infty} \varepsilon \left(\frac{z}{2\eta_0} + \tau_0 \xi_1 \right) \text{sech}^2 z \, dz \right] \right\rangle \tag{4.6.9}$$

For a δ-correlated Gaussian process with a zero mean,

$$\langle \varepsilon \rangle = 0 \ , \quad \langle \varepsilon(\tau_1, \xi_1)\varepsilon(\tau_2, \xi_2) \rangle = 2\sigma^2 \delta(\tau_1 - \tau_2)\delta(\xi_1 - \xi_2) \ ,$$

we obtain

$$\langle \eta^2 \rangle = \eta_0^2 \exp \left\{ 4\xi \left[\frac{8}{3} \eta_0 \sigma^2 (1 - \Gamma_1 \xi) - \Gamma \right] \right\} \ , \tag{4.6.10}$$

implying that, for $2\sigma^2 > 3/4(\Gamma/\eta_0)(1 - \Gamma_1 \xi) \approx 3\Gamma/4\eta_0$, a stochastic resonance will appear. Note also that, in the case where the random function depends only on the coordinate τ, (4.6.10) becomes

$$\langle \eta^2 \rangle = \eta_0^2 \exp \left\{ 4\xi \left[\frac{8}{3} \eta_0 \sigma^2 \xi (1 - \Gamma_1 \xi) - \Gamma \right] \right\} \ . \tag{4.6.11}$$

If ε is only a function of ξ, then

$$\langle \eta^2 \rangle = \eta_0^2 \exp \left\{ 4\xi \left[\sigma^2 (1 - \Gamma_1 \xi) - \Gamma \right] \right\} \ . \tag{4.6.12}$$

These results are readily extended to the case $\varepsilon \neq 0$. When $2\langle \varepsilon \rangle >> (\langle \varepsilon^2 \rangle - \langle \varepsilon \rangle^2)$, we have

$$\langle \eta^2 \rangle = \eta_0^2 \exp \left\{ 4\xi \left[\langle \varepsilon \rangle \left(1 - \frac{\Gamma_1 \xi}{2} \right) - \Gamma \right] \right\} \ . \tag{4.6.13}$$

Equations (4.6.10–13) indicate that there exists a range of parameter values where a stochastic parametric resonance of the solitons occur.

Gordon and *Haus* (1986) showed that amplifier noise leads to a random shift of the soliton carrier frequency, and to a random variation of its velocity. The effects of small random perturbation on the soliton were considered. The noise field has the form

$$\delta q = i a q_s \tanh(A\xi - \tau) \, ,$$

where a is an amplitude. This term perturbs the frequency $\Omega_0 = \omega - \omega_0$. The analysis showed that the mean-square velocity shift is equal to $\delta v_g^{-1} = -\delta\Omega$, and

$$(\delta\Omega)^2 = (G - 1)\frac{A}{3N_0} \, ,$$

where G is the coherent amplifier gain, N_0 is the number of photons per unit energy, and A is the soliton amplitude. The possibility of suppressing the Gordon-Haus effect by a suitable design of the frequency-dependent gain was mentioned by *Kodama* and *Hasegawa* (1991).

The influence of random coupling on solitons in NLDC has been studied by *Darmanyan* (1992).

4.7 Dynamic Chaos of Optical Solitons

In addition to the previous mechanisms of stochastic influences on optical solitons, there exists also a so-called "dynamic chaos". Dynamic chaos is understood to be a phenomenon such as random (unpredictable) motion of a fully deterministic system, whose equations of motion do not contain any random parameters or any noise. For example, dynamic chaos is observed in a nonlinear wave system when unstable modes are in the presence of periodic, dissipative, or otherwise nonintegrable perturbations.

Nonlinear dynamics is a much studied subject (*Chirikov*, 1979; *Lichtenberg* and *Lieberman*, 1981; *Zaslavsky*, 1984). Dynamic chaos cannot be accounted for by random initial conditions of the problem or random time-varying parameters of the system. The main reason for this is a strong local instability of the motion in the phase space resulting in stochastic trajectories. The theory of dynamic stochasticity is well advanced for Hamiltonian systems; in particular, there is a systematic formalism based on a resonance perturbation theory, where one investigates the splitting of separatrices and the formation of stochastic layers near them (Chirikov's criterion, Melnikov's method). The theory of dissipative dynamic systems is also considerably developed (*Lichtenberg* and *Lieberman*, 1981).

In the case of nonlinear extended systems, the situation is more complicated. At present, there are no exact analytic criteria available to allow the determination of conditions for dynamic turbulence to evolve in a nonlinear wave system. Major results in this case are obtained by numerical simulation (*Bishop* et al., 1983; *Nozaki* and *Bekki*, 1983).

Only in those cases when the nonlinear waves are described by equations which are close to integrable ones, can qualitative results be obtained. The procedures applied for this purpose could be the following:

a) localized modes (solitons, breathers) or nonlinear periodic modes are separated;

b) using perturbation theory for nearly integrable systems, a set of equations for mode parameters is derived so that dynamics of an infinite-dimensional system is reduced to a finite-dimensional one;

c) the qualitative methods of dynamical systems are applied to the resulting finite-dimensional problem (*Chirikov*, 1979; *Lichtenberg* and *Lieberman*, 1981) and the conditions for the occurrence of stochasticity are defined.

Such an analysis was carried out by *Bishop* et al. (1983), *Nozaki* and *Bekki* (1983), *Abdullaev* (1983, 1989), *Abdullaev* et al. (1985), and *Aranson et al* (1984) for solitons and breathers in the NLS, SG and Zakharov systems, and the forced Toda lattice, among others.

From the viewpoint of the occurrence of dynamic chaos of optical solitons the investigations by Abdullaev et al. and Aranson et al. are of particular interest. *Abdullaev* (1983) and *Aranson* et al. (1984) determined the conditions for the occurrence of chaos in envelope soliton dynamics under the action of a broad class of periodic perturbations; *Abdullaev* et al. (1985) defined dynamic chaos of SG breathers under the action of a parametrically acting periodic field. These results are important for the determination of chaotic propagation of optical solitons along a fibre, and for short pulses in a resonant medium (self-induced transparency).

Later, we will analyze the chaotic dynamics of optical solitons under the action of an external discrete perturbation caused by modulation of the medium interface (*Pikovsky*, 1985), and by a periodic modulation of the medium parameters (*Abdullaev* and *Umarov*, 1985).

Let a plane beam propagate in a nonlinear medium. Its propagation is described by the NLS equation in terms of a complex electromagnetic field amplitude. In dimensionless variables the equation has the form

$$i\frac{\partial u}{\partial z} + \frac{\partial^2 u}{\partial x^2} + 2|u|u^2 = 0 \,, \tag{4.7.1}$$

where z and x are the coordinates along the direction of beam propagation and transverse to it, respectively.

We now study the motion of a soliton beam in a modulated waveguide formed by ideally reflecting walls ($-\infty < p < \infty$, $\cos \kappa p < g < L$) (Fig. 4.6). Assume first that the beam is narrow, so that its dynamics for $a << L$ reduces to a mapping:

$$\varphi_{n+1} = \varphi_n - 2a\kappa \sin \kappa p_n \, ,$$

$$p_{n+1} = \frac{2L}{tg\varphi_{n+1}} + p_n \, . \tag{4.7.2}$$

Let the condition $\nu = a\kappa \ll 1$ be fulfilled, i.e., each reflection from the wall gives a small change in the angle φ. Then, from the second equation in (4.7.2), we obtain a first approximation in ν, and

$$p_{n+1} \simeq \frac{2L}{tg\varphi_n} + \frac{4L\nu}{\sin^2 \varphi_n} \sin \kappa p_n + p_n \, . \tag{4.7.3}$$

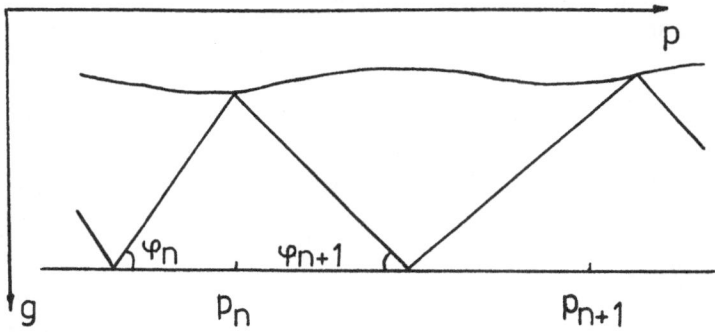

Fig. 4.6. Motion of a soliton beam in a waveguide, p is parallel to the waveguide axis, g is parallel to the cross sectional plane

Let us analyze the system (4.7.2–3) with the help of the criteria of stochasticity of two-dimensional mapping, which has been considered by *Sinai* (1966). Its application shows that stochasticity exists when $|dp_{n+1}/dp_n| > 1$, i.e., in the presence of phase p_n dilatation. This provides a criterion of chaos which is fully analogous to the condition of appearance of chaotic motion of a billiard particle

$$K = \nu\kappa L \gg 1 \, . \tag{4.7.4}$$

Provided (4.7.4) is fulfilled within a broad range of φ_n, the beam dynamics will be chaotic.

Let us now regard the consequences of a finite size for the soliton. The reflection comes from a curved surface whose local shape near the point p_0 has the form

$$y(p_0 + \Delta p) = a \cos \kappa p_0 - \Delta p \kappa \sin \kappa p_0 + \Delta p^2 \frac{a\kappa^2}{2} \cos \kappa p_0 \, . \tag{4.7.5}$$

The beam reflection (4.7.2) from the surface (4.7.5) results in a distortion of the field phase

$$u_{\text{ref}}(x) = u_0(x) \exp(i\varepsilon x^2) \, , \tag{4.7.6}$$

where $\varepsilon = 0.5k_0a\kappa^2 \cos \kappa p_0 \sin^{-1} \varphi_n$, and k_0 is the wave number of the beam field.

Let $u_0(x)$ be an exact soliton solution of (4.7.1). Since the field $u_{\mathrm{ref}}(x)$ is not an exact soliton solution, it has, in addition to a soliton, a nonsoliton perturbation which spreads in space.

Provided the reflected field is not too different from the incident one, i.e., for $|\varepsilon\eta^{-2}| << 1$, the parameters of a reflected soliton can be defined by the IST-based perturbation method. The result thus obtained shows, in particular, that an envelope soliton propagating in a resonator with a periodically moving wall decays randomly.

Next, we will show that dynamic chaos might occur in the problem of an NLS soliton propagating over a nonstationary medium (*Abdullaev* and *Umarov*. 1985):

$$iq_t + \frac{1}{2}q_{xx} + |q|^2q + \frac{1}{2}\varepsilon f(x)\phi(t)q = 0 , \tag{4.7.7}$$

where the functions $f(x)$ and $\phi(t)$ describe medium inhomogeneity and nonstationarity, respectively. If we proceed with the variables

$$q = \frac{E}{E_0} , \quad t = \frac{y}{2k} ,$$
$$E_0 = \frac{1}{2}\left(\frac{n_0}{n_2}\right)^2 , \tag{4.7.8}$$

where x and y are longitudinal and transverse coordinates, respectively, we can easily see that (4.7.7) describe the propagation of an intense plane electromagnetic beam in an inhomogeneous medium. Consider the evolution of the one-soliton solution q_s of (4.7.7). Perturbation theory for solitons is valid for times satisfying

$$t_s << t \le \varepsilon^{-1} , \quad t_s \sim \frac{\ell_s}{v_s} ,$$

where ℓ_s is a soliton width.

Equations for the soliton parameters η, μ, δ, and ζ are (for the case $f(x) = \sin \lambda x$)

$$\frac{d\eta}{dt} = 0 , \quad \eta = \eta_0 , \quad \frac{d\zeta}{dt} = \mu , \tag{4.7.9}$$

$$\frac{d\mu}{dt} = -\frac{\varepsilon\pi\lambda^2}{8\eta_0 \sinh(\pi\lambda/4\eta_0)}\phi(t) \cos \lambda\xi . \tag{4.7.10}$$

From these, we derive a single equation for the soliton centre coordinate:

$$\frac{d^2\zeta}{dt^2} + a(1 + \varepsilon_1 \sin \alpha t) \sin \zeta = 0 ,$$
$$a = \frac{\varepsilon\pi\lambda^3}{8\eta_0 \sinh(\pi\lambda/4\eta_0)} . \tag{4.7.11}$$

It is assumed here that $\phi(t) = 1 + \varepsilon_1 \sin \alpha t$. This equation is similar to that describing oscillations of a parametrically driven pendulum. It can be analyzed using appropriate criteria for the overlapping of nonlinear resonances (Chirikov's criterion, for example).

For small ζ and $\alpha/a = 2$, a parametric resonance occurs. An analysis reveals that in the vicitnity of the separatrix of the unperturbed motion a stochastic layer forms, causing random motion. The condition for the occurrence of stochasticity is

$$K = \frac{4\pi\varepsilon_1}{\omega_r} \left(\frac{\alpha}{a}\right)^3 \exp\left(-\frac{\pi\alpha}{2a}\right) \geq 1 , \qquad (4.7.12)$$

where

$$\omega_r = 32 \exp\left(-\frac{2\pi r a}{\alpha}\right) ,$$

and r is an integer. It is seen that for any $\alpha/a \to \infty$, $K \geq 1$ in a vicinity close to the separatrix.

Of interest is the fine structure in the transition to a stochastic behaviour. A numerical analysis (*McLaughlin*, 1981) shows that for $\varepsilon_1 = 0.6$, $\alpha = 2$, stochastic dynamics occurs in the system. Increasing α from 0 to $\alpha_{\text{crit}} = 0.6$, a period-doubling array occurs.

We note also that chaotic dynamics of solitons in a fibre can be observed when there is a weak electromagnetic wave present in a different mode of the fibre. In this case the perturbation term in the NLS equation is

$$\varepsilon(\xi,\tau) = \sum \varepsilon_i \cos(k_i\xi - \Omega_i\tau)q(\xi,\tau) ,$$

i.e., the form of a wave packet with the spectral width $\Delta\Omega$. Chaotic motion of the NLS soliton in a wave packet field has been studied by *Abdullaev* (1983) and by *Bass* et al. (1989). Using the Chirikov's criterion, we obtain the next estimate for the appearance of stochasticity:

$$K = \frac{\Omega^2\varepsilon\alpha}{8(\Delta\Omega)^2\sinh(\alpha)|2\mu - v_g^{-1}|^2} \geq 1 , \quad v_g = \frac{d\Omega}{dk} , \quad \alpha = \frac{\pi\Omega}{4\nu} .$$

Close to resonances this condition takes the simple form $K \approx \Omega^2/8(\Delta\Omega)^2 > 1$. Chaotic behaviour OS is also possible in an amplifying medium (*Kodama*, 1985).

4.8 Optical Turbulence in Passive Optical Resonators

We proceed now with the discussion of soliton dynamics in a bistable optical ring resonator (*Moloney*, 1985). An optical ring resonator serves as a convenient model frequently applied to the analysis of optical bistability. The ring resonator is simpler than a Fabry-Perot interferometer, because of the presence of standing waves which neccessarily occur in the latter. The ring resonator contains a two-level nonlinear medium with saturable nonlinearity, and has input and output mirrors with reflection coefficient R. Assuming the relaxation time to be less than the round trip time of the cavity, the appropriate evolution equation is

$$2i \frac{\partial}{\partial \zeta} G_n + \Delta_{tr} G_n - \left(\frac{G_n}{1 + 2|G_n|^2} \right) = 0 \,, \tag{4.8.1}$$

with the boundary conditions

$$G_n(x, y, 0) = a(x, y) + \mathrm{Re}^{ikL} G_{n-1}(x, y, p) \,, \tag{4.8.2}$$

$$n > 0 \,, \quad G_0 = 0 \,;$$

here x, y are normalized values for the transverse coordinates. In the case of a Kerr-type nonlinear medium (corresponding to the limit $G << 1$), the above becomes a two-dimensional NLS equation. Here G_n is the amplitude of the field inside the resonator, n is the number of round trips over the resonator ring, Δ_{tr} is a transverse Laplacian. Equation (4.8.2) is derived from the boundary conditions for the resonator. Further, $\alpha_0 L_1$ is the linear absorption coefficient per a pass; $\Delta = (\omega_{ab} - \omega)/\gamma_{\perp}$ is a dimensionless frequency detuning from atomic resonance, where $\gamma_{\perp}$ is the dipole relaxation time, and $F = n_0 \omega_{ab}/\lambda L$ is the Fresnel number which measures the influence of diffraction in linear propagation. The parameter $pL_1 = \alpha_0 L_1/\Delta$ measures the influence of diffraction. Equations (4.8.1–2) are solved as follows.

1) First the external pump is taken in the form

$$a(x, y) = a_p \exp[-(x^2 + y^2)] \,,$$

then

$$G_1(x, y, 0) = a(x, y) \,, \quad G_0 = 0 \,.$$

2) Values $G_1(x, y, 0)$ are taken as the initial data for (4.8.1), which is then solved to determine $G_1(x, y, p)$.

3) The values of $G_2(x, y, 0)$ are then found from the resulting $G_1(x, y, p)$ with the use of (4.8.2). This procedure is repeated until an asymptotic state is reached.

Note that if the ring resonator is "straightened" into a line, we have the problem of the propagation of a laser beam with periodical loss. Two types of asymptotic states are possible, depending on the resonator quality. In a high-finesse resonator an asymptotic state will be largely defined by soliton solutions of a NLS equation, so that external pumping can be considered to be a weak perturbation. Solitons in this case represent transversal structures (rings). Soliton parameters are defined by fixed points of a certain two-dimensional mapping. Conversely, when dissipative effects are large (low-finesse resonator case), solitonic and other nonlinear effects are unimportant. This limit is described by the plane-wave map.

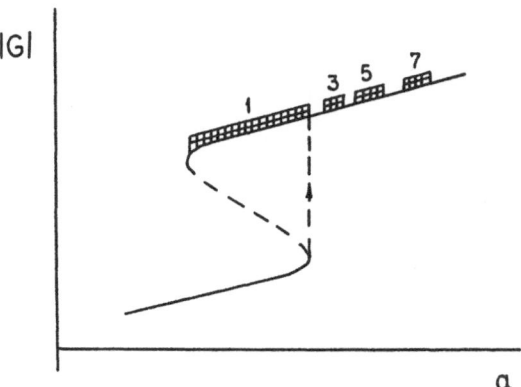

Fig. 4.7. A bistable loop showing the disposition of region where transverse n-solitary wavetrains are stationary asymptotic states. Cross-hatches indicate $n = 1, 3, 5, 7$ soliton states. From *Moloney* (1985)

Let us analyze a one-dimensional transverse case. As numerical simulations show, the propagation of a laser beam in a nonlinear resonator yields bistability. Figure 4.7 represents a bistable loop in $(|G| - a)$-plane showing the relative disposition of regions on the high-transmission branch where transverse n-solitary wave-trains are stationary asymptotic states. The region, corresponding to 1, 3, 5, and 7 solitons is marked by the cross hatches. Between these shadow regions, slow recurrent oscillations occur in time between even and closest odd-numbered wavetrains.

If we follow the upper branch of the hysteresis loop to the left, we see a decrease of the soliton amplitude and the increase of its width.

The dynamic evolution of the laser beam is presented in Fig. 4.8. There are three stages of evolution:

a) the intial evolution when a beam traverses the resonator for the first time;
b) the formation of a pedestal at the edges and flattening of field shape at the centre of beam;
c) seven solitons are formed after 200 runs. After 20 runs the profile is sharpened, after 30 runs a train of solitons appears at the edges. For $a = 0.0375$, seven solitons are formed and, in the vicinity of the single-soliton formation, hysteresis is observed.

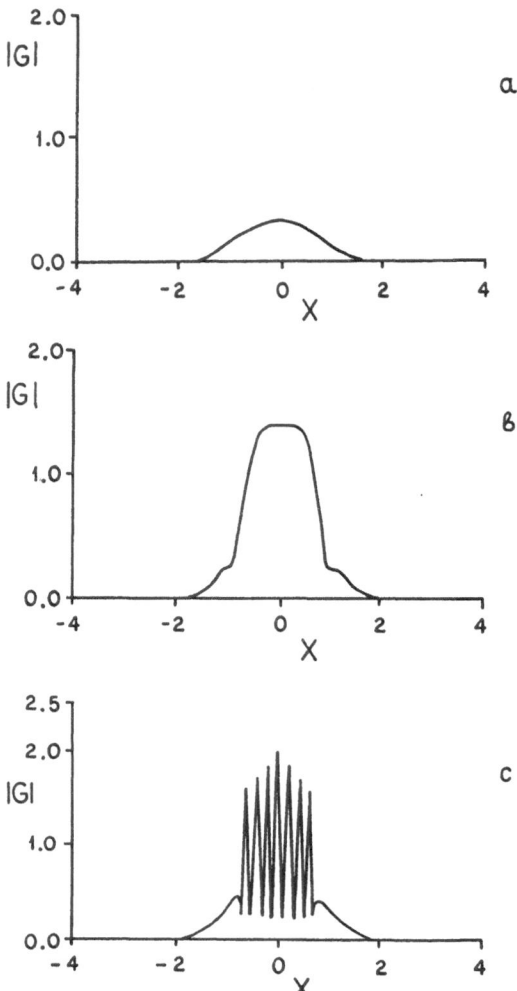

Fig. 4.8a–c. Three stages in the dynamic switching of a one-dimensional transverse beam: (a) is the initial Gaussian pump, (b) the state after the 20th pass, (c) is the seven-soliton state after 200 resonator passes. From *Moloney* (1985)

Using integral invariants (4.8.1), one can evaluate the number of solitons generated by a pump wave. It can be shown that the half-widths of soliton waves are $\sim 1\sqrt{f}$, $f = (4\pi F/\ln r)p$. A two-dimensional case can be analyzed by numerical simulation. It shows that, unlike the one-dimensional case where a finite state is formed as n-soliton arrays, the central part of the beam exhibits oscillations after 100 runs over the resonator. For strongly saturated values of amplitudes stable circular waves (solitons) are separated.

The simulation of $|G_n|^2$ showed that the final asymptotic state is a stable period-2 oscillation between transverse spatial rings. As $|G|^2$ is growing, the period-2 oscillation is destabilized and modulated. When driven harder, the modulation period decreases and the traces become more chaotic. There is evidence for the existence of soliton-like spatial structure, even when the beam is undergoing chaotic temporal motion (*Moloney*, 1986).

Interesting phenomena occur when an optic fibre is chosen as a nonlinear resonator medium (*Blow* and *Doran*, 1984). If a periodic pulse array

$$q_{\text{pump}}(\tau, \xi = 0) = A \sum_n \text{sech}(\tau - n\tau_0)$$

is launched into a fibre with loss. Then, according to numerical calculations with increasing signal amplitude A, a period-doubling bifurcation array is formed: for $A = 1.526$ and $A = 1.89$. For $A \rightarrow 1.98$, one observes chaotic behaviour of the wave field in time.

The observed complex dynamic and chaotic behaviour of solitons in resonators is of significance for developments dealing with fabrication of soliton lasers (see Sect. 2.10).

5. Optical Solitons in Resonant and Active Media

In order to investigate the propagation of short pulses in a medium with resonant impurities, it is necessary to take the medium's dipole character into account. Here, new phenomena unknown in media possessing only a Kerr-type nonlinearity are possible. In this chapter, we will study solitonic phenomena in both active and passive resonant media. We start with some general problems of wave propagation in resonant media, derive the coupled Maxwell-Bloch equations, then investigate some problems of random pulse propagation. In such systems, a new type of optical solitary waves – autosolitons – are possible. We then study phenomena in the generalized Maxwell-Bloch system , and include the effects of Kerr-type nonlinearity and waveguide structures. We finally discuss optical soliton generation in SRS as possible means of generating short intense pulses.

5.1 The Maxwell-Bloch System. Soliton Solutions

Consider the problem of pulse propagation in a resonant medium. It is useful to represent a medium as a system of two-level atoms interacting with the radiation field. Let the frequency of the allowed atomic transition be ω_0, the frequency of the electromagnetic wave be ω, and assume that ω is close to ω_0. We now derive a system of equations describing one-dimensional propagation of the electromagnetic field along the z-axis. Here, we follow works by *Allen* and *Eberly* (1975), and *Lamb* (1980).

A wave propagating in a dipole medium interacts with a set of two-level atoms. Physically, the interaction is similar to the case when a particle with spin 1/2 interacts with a magnetic field. The evolution is described by the optical Bloch equations which couple the population inversion w with the atomic dipole moment envelopes u and v (respectively, the in-phase and phase quadrature components relative to the electromagnetic field E):

$$\dot{w} = -\kappa \mathcal{E} v - \frac{w}{T_1} \, ,$$

$$\dot{u} = -\Delta u - \frac{u}{T_2} \, , \tag{5.1.1}$$

$$\dot{v} = \Delta u - \frac{v}{T_2} + \kappa \mathcal{E} w ;$$

$$\kappa = \frac{2d}{\hbar} .$$

Here $\mathcal{E}$ is the envelope of the electromagnetic field; $\Delta = \omega_0 - \omega$ is the frequency detuning; d is the value of the matrix dipole element, T_1 is a longitudinal decay time describing the damping of the inversion and T_2 is the transverse damping time which describes incoherent contributions with loss. It is connected, for example, with local random strictions in solid states or with phase-distorting collisions.

The propagation of the electromagnetic field along z axis is described by the wave equation

$$c^2 \bar{E}_{tt} - \bar{E}_{zz} = \frac{4\pi}{c^2} P_{tt}(z, t) . \tag{5.1.2}$$

The polarization density is

$$p = n_0 e \, x(t, z) = n_0 d \langle \sigma_1(z, t) \rangle ,$$

where $e \, x(t, z)$ is a dipole moment of atom, n_0 is the atomic density, and σ_1 is the Pauli matrix. The brackets $\langle \ldots \rangle$ denote the averaging on two-level atom states. Taking the fields $\bar{E}$ and P in the form

$$\bar{E} = \mathcal{E}(t, z) \left[e^{i(\omega t - K z)} + \text{c.c.} \right] ,$$

$$P = n_0 d \int g(\Delta') \left[u \cos(\omega t - K z) - v \sin(\omega t - K z) \right] d\Delta' , \tag{5.1.3}$$

where $g(\Delta)$ is a distribution function for the frequency detuning. Substituting these expressions into (5.1.2), we obtain

$$(K^2 - k^2)\mathcal{E}(t, z) = 2\pi k^2 n_0 d \int_{-\infty}^{\infty} u(t, z; \Delta') g(\Delta') d\Delta' ,$$

$$2 \left[K \frac{\partial}{\partial z} + \frac{k}{c} \frac{\partial}{\partial t} \right] \mathcal{E}(t, z) = 2\pi k^2 n_0 d \int_{-\infty}^{\infty} v(t, z; \Delta') g(\Delta') d\Delta' , \tag{5.1.4}$$

where $k = \omega/c$.

These, together with the Bloch equations (5.5.1), are the Maxwell-Bloch (MB) equations. If we neglect damping, then the equality

$$w^2 + u^2 + v^2 = 1 .$$

is valid. We now introduce the function $\theta(z, t)$, where

$$\theta = \kappa \int_{-\infty}^{t} \mathcal{E}(t', z) dt' . \tag{5.1.5}$$

It is also useful to introduce the variable $v(z, t; 0) = -\sin\theta(z, t)$, and to assume that the factorization $v(z, t; \Delta) = v(z, t; 0)F(\Delta)$ is valid. Here, $F(\Delta)$ is the "spectral response" function. As a result, we obtain the equation for $\theta(z, t)$

$$\ddot{\theta} - \frac{1}{\tau_0^2}\sin\theta = 0 , \qquad (5.1.6)$$

where

$$\frac{1}{\tau_0^2} = \frac{\Delta^2 F(\Delta)}{1 - F(\Delta)} .$$

This is the pendulum equation, which has the solution

$$\theta = 4\tan^{-1}\left[\exp\frac{t - t_0}{\tau_0}\right] ,$$

$$\mathcal{E} = \frac{2}{\kappa\tau}\,\text{sech}\left[\frac{t - t_0}{\tau_0}\right] . \qquad (5.1.7)$$

The solution describes a self-induced transparency regime, where an intense pulse propagates through a resonant medium without changing its shape (*McCall* and *Hahn*, 1967).

The most important property of the reduced Maxwell-Bloch system is its complete integrability (*Lamb*, 1980). Following Lamb, we introduce new variables

$$\lambda = (v + iu)\exp(i\varphi) ,$$

$$E = \frac{d}{h\Omega}\bar{E}\exp(i\varphi) , \qquad (5.1.8)$$

$$\eta = \frac{\Delta}{2\Omega} , \quad \Omega^2 = \frac{2\pi n_0\omega_0 d^2}{h} .$$

Substituting these into (5.1.1,4), and neglecting the damping effects, we obtain

$$\frac{\partial E}{\partial\xi} = \langle\lambda\rangle , \qquad (5.1.9a)$$

$$\lambda_\tau + 2i\eta\lambda = EN , \quad \lambda_\tau^* - 2i\eta\lambda = E^*N , \qquad (5.1.9b)$$

$$N_\tau = -\frac{1}{2}\left[E^*\lambda + E\lambda^*\right] , \qquad (5.1.9c)$$

where $\xi = \Omega z/c$, $\tau = \Omega(t - z/c)$.

The single soliton solution is

$$E = 4\nu e^{-i(\alpha_1\xi + 2\mu\tau + \theta)}\text{sech}(\nu\xi - 2\nu\tau + \gamma) ,$$

where $\alpha_1 + i\beta_1 = 1/2\langle\langle 1/(\eta - k_1 - i\varepsilon)\rangle\rangle$, $k_1 = \mu + i\nu$, θ is the initial phase, γ defines the soliton centre, and $\langle\langle \ldots \rangle\rangle$ denotes an average over frequency detuning.

There are many applications of the MB equations to different physical problems, such as to self-induced transparency (*Lamb*, 1980), self-focusing of resonance pulses, and superfluoresence (*Gabitov* et al., 1985).

We now describe results of soliton generation in a resonant medium where these are generated from a localized initial disturbance having a phase modulation. This problem was considered by *Elyutin* et al. (1988), who used the inverse scattering transform. The initial profile is taken to be

$$E(0, t) = E_0 \text{sech} \frac{\tau}{\tau_0} \exp[i\varphi(t)] \, ,$$

where the amplitude E_0 is chosen so that the integrated square of the pulse envelope is either 2π or 4π. The Zakharov-Shabat system for (5.1.9) is

$$v_{1,\tau} + i\zeta v_1 = \frac{i}{2} E(\xi, \tau) v_2 \, ,$$
$$v_{2,\tau} + i\zeta v_2 = -\frac{i}{2} E^*(\xi, \tau) v_1 \, ,$$
(5.1.10)

It is assumed that the phase $\varphi(t)$ is a Gaussian delta-correlated random process,

$$\langle \varphi \rangle = 0 \, , \quad \langle \varphi(t)\varphi(t') \rangle = \Gamma^2 \delta(t - t') \, .$$

Computer simulation indicates that there exists a minimum critical value of Γ_c, starting from which the solitons are formed. One can estimate this critical value by applying the random phase approximation as used in the theory of disordered systems (*Ziman*, 1979).

This approximation corresponds to the replacement of $\mathcal{E}(z, t)$ by its mean value $\langle E(\xi, \tau) \rangle$ in the Zakharov-Shabat system. Here, the angular brackets denote an ensemble-average rather that integration over frequency, as before:

$$\langle E(\xi, \tau) \rangle = \text{sech}\tau \exp\left(-\frac{\Gamma^2}{2}\right) = A \, \text{sech}\tau \, .$$
(5.1.11)

The eigenvalue ζ_1 is defined by the equality $\zeta_1 = i(A - 1/2)$. Thus,

$$\zeta_1 = i\left[\exp\left(-\frac{\Gamma^2}{2}\right) - \frac{1}{2}\right] \, ,$$
(5.1.12)

and so the corresponding critical point Γ_c^1 is

$$\Gamma_c^1 = 2\ln 2 = 1.18 \, .$$
(5.1.13)

This value is in accordance with data from the computer simulation. In the case of a large soliton number, the estimate gives

$$\zeta_n(\Gamma) = i\left[E_0 \exp\left(-\frac{\Gamma^2}{2}\right) - n + \frac{1}{2}\right] \, .$$
(5.1.14)

The corresponding critical values $\Gamma_c^{(k)}$ are

$$\Gamma_c^{(k)} = \left[2\ln\left(\frac{2l_0}{2k-1}\right) \right]^{1/2} .$$

Hence, the increase of random phase modulation Γ results in a decrease of the number of 2π pulses formed.

The generalized Maxwell-Bloch sytem has recently been used by *Gabitov* et al. (1991) to investigate the amplification of femtosecond solitons in erbium-doped fibers.

5.2 Optical Solitons in an Active Fibre

The propagation of short pulses in a real fibre demands that proper account be taken of impurities, which usually promote radiation absorption. However, there are situations where the impurities are of resonant character. The dynamics of optical solitons in fibres containing such resonant impurities is unusual, and gives rise to the occurrence of new types of solitons and autosolitons. A stationary solitary wave existing in a medium with amplification and dissipation is called an autosoliton; its parameters are defined by the medium parameters and do not depend on the initial pulse parameters.

In this section, we will study soliton propagation in a single mode fibre with two-level atoms as resonant impurities (*Maimistov and Manykin*, 1983). The problem is described by the Maxwell-Bloch equations, to which the terms which take account of the influence of Kerr nonlinearity and dispersion are added:

$$iE_Z + aE_{\tau\tau} + g|E|^2 E + \langle \varrho \rangle = 0 , \tag{5.2.1}$$

$$\varrho_\tau = i\delta\varrho + ifEN , \tag{5.2.2}$$

$$N_\tau = 2if(\varrho E^* - \varrho^* E) . \tag{5.2.3}$$

Here, $Z = z/\ell_s$; $\tau = t - z/v_g$; $a = \ell_s/z_d$; $g = \ell_s/z_{nl}$; where ℓ_s is a characteristic length for resonant absorption, ℓ_{nl} is a nonlinear length, z_d is a dispersive length, τ_0 is the pulse duration, and A_0 is a maximum amplitude. The quantity $f = \bar{d}A_0\tau_0 h^{-1}$ describes the character of interaction between a propagating field and the two-level atoms, where $\bar{d}$ is an effective matrix element of the dipole transition between the resonant states. We want to know whether soliton solutions of (5.2.1,3) exist. *Maimistov* and *Manykin* (1983), applying the IST, showed that such states exist only at certain parameter values. We represent this system in the Lax form:

$$L_Z = A + [A, L] , \tag{5.2.4}$$

where

$$L = \begin{pmatrix} -i\lambda, & \alpha_1 E \\ \alpha_2 E^*, & i\lambda \end{pmatrix} , \quad A = \begin{pmatrix} A, & B \\ C, & -A \end{pmatrix} . \tag{5.2.5}$$

Here, λ is a spectral parameter. The constants α_1, α_2, A, B, C are chosen so that (5.2.4) coincides with the set (5.2.1,3). Equation (5.2.4) implies

$$A_\tau = E_1 C - E^* B , \tag{5.2.6}$$

$$B_\tau + 2i\lambda B = \alpha_1 E_Z - 2\alpha_1 E A , \tag{5.2.7}$$

$$C_\tau - 2i\lambda C = \alpha_2 E_Z^* + 2\alpha_2 E^* A . \tag{5.2.8}$$

Representing B and C as linear combinations of ϱ, E, E_τ and imposing the compatibility condition, one can infer that soliton solutions exist provided the conditions

$$g = -2\alpha_1\alpha_2 a , \quad f^2 = -\alpha_1\alpha_2 , \tag{5.2.9}$$

are met, i.e.,

$$g = 2af^2 . \tag{5.2.10}$$

In terms of dimensional variables, we have

$$\frac{\ell_d}{\ell_{n\ell}} = 2f^2 , \tag{5.2.11}$$

i.e.,

$$\frac{n_0 n_2}{2} \left(\frac{\hbar\omega}{c\bar{d}}\right)^2 = \left|\frac{\partial^2 k}{\partial\omega^2}\right| . \tag{5.2.12}$$

Thus, the analysis shows that there exists an optical soliton which is simultaneously a 2π-pulse of self-induced transparency and a soliton to the NLS equation. This indicates that a part of the waveguide with resonant impurities where a population inversion has been produced by optical pumping can serve as an active element. In this case, the set (5.1.1–3) will include the difference of resonance level populations with the opposite sign, thus resulting in gain.

Note that the condition (5.2.9) for the existence of a soliton in a single-mode fibre containing resonant impurities are rigid and are within the threshold for breakdown of the fibre. For this reason, corresponding soliton-like pulses are possible in waveguides which have the largest values of n_2.

Consider the propagation of an optical pulse in a fibre activated by resonant impurities. As was shown by *Grigoryan* et al. (1989), the evolution of a pulse in a weakly dispersive inverted medium with a Kerr nonlinearity and linear losses is described by the perturbed NLS equation, including effects of weak saturation of gain:

$$iq_\xi + \frac{1}{2} q_{\tau\tau} + |q|^2 q = -i\tilde{\alpha}q \int_{-\infty}^{\tau} |q|^2 d\tau' + i\tilde{\beta}q . \tag{5.2.13}$$

The coefficients $\tilde{\beta}$ and $\tilde{\alpha}$ represent the linear gain and the gain saturation; $\tilde{\alpha}$ is proportional to the density of amplifying centres. From (5.2.13), an equation for the integral

$$N = \int_{-\infty}^{\infty} |q|^2 d\tau$$

is obtained. If $|q| \rightarrow 0$ at $|\tau| \rightarrow \infty$, then we have

$$N_\xi = 2\beta N - \tilde{\alpha} N^2 . \tag{5.2.14}$$

Substituting into this the single soliton from (2.3.17), we obtain an equation for the soliton amplitude

$$\frac{d\eta}{d\xi} = 2\eta(\tilde{\beta} - \tilde{\alpha}\eta) . \tag{5.2.15}$$

It is seen that at $\eta_c = \tilde{\beta}/\tilde{\alpha}$ there exists a steady-state (an autosoliton) in the medium where the energy growth rate is fully compensated by the energy losses. The autosoliton has the form

$$q(\xi, \tau) = \frac{\tilde{\beta}}{\tilde{\alpha}} \exp\left\{ i\left[\zeta_0\tau + \frac{1}{2}\zeta_0^2 - \frac{\tilde{\beta}^2}{\alpha^2}\tilde{\xi} \right] \right\}$$

$$\times \operatorname{sech}\left[\frac{\tilde{\beta}}{\tilde{\alpha}}(\tau + \zeta_0\xi + \tilde{\alpha}\xi) \right] . \tag{5.2.16}$$

Numerical calculations showed that this solution is unstable when $\tilde{\alpha}, \tilde{\beta} > 1$. The asymptotic form is independent of the initial pulse energy.

Predicted effects can be observed in a single-mode fibre doped with Nd ions. When the population inversion density of ions is 10^{19} cm^{-3}, $k'' = -1.1 \times 10^{-26}$ s^2/cm and the dispersion length is equal to 100 m (for 10-ps pulse duration). Parameters $\tilde{\alpha}$ and $\tilde{\beta}$ are ~ 5 for a damping coefficient ≈ 0.27 sm^{-1}. Then, the theory predicts soliton generation with a soliton duration $\tau_b = 12$ ps and a power density of ≈ 42 mW/cm^2.

The transmission of dark solitons in optical fibers with amplification was investigated numerically by *Hammaide* et al. (1991).

5.3 Scattering of a Weak Wave on an Optical Soliton

In this section, we will consider the effects caused by the propagation of two types of waves: weak and intense. If the intensity of one wave is large, there results a strong change in the state of the medium, while propagating in the medium the soliton causes local changes of the dielectric permeability of this medium. This localized region of the medium with the modified dielectric permeability propagates with a certain high velocity close to the light velocity. Thus, we have a problem of wave-scattering in a medium with a nonstationary moving inhomogeneity.

A similar problem in which a weak wave scatters from a shock front was examined by *Ostrovsky* (1959). He predicted the possibility of a change in the frequency of the weak wave. Application of this effect to the optical soliton case is important, because here the value of the Doppler effect is large ($v_s \sim c_0$).

This application was treated by *Rupasov* (1982), and by *Bolshov and Likhansky* (1985) for the case of a three-level medium. We will follow the ideas developed by Rupasov.

Consider a gas of three-level atoms, and assume that the carrier frequency of a strong pulse is near-resonant with the transition $1 \rightarrow 2$, and that the frequency ω of the weak field is similarly near-resonant with the transition $2 \rightarrow 3$. Moreover, it will be assumed that in the absence of the field E_1 only the level 1 is occupied, the levels 2 and 3 being vacant, so that the medium is transparent to the second wave.

The Hamiltonian for the three-level system interacting with the fields E_1 and E_2 is

$$H = H_0(t) + H_{\text{int}}(t) ,$$

$$H_0(t) = \sum_{i=1}^{3} \mathcal{E}_i c_i^+ c_i - \frac{1}{2} d_1 \left(E_1 c_2^+ c_1 + E_1^* c_1^+ c_2 \right) , \qquad (5.3.1)$$

$$H_{\text{int}}(t) = -\frac{1}{2} d_2 \left(E_2 c_3^+ c_2 + E_2^* c_2^+ c_3 \right) .$$

Here, c_i^+, c_i are Fermi operators for electron creation and annihilation at the level $i = 1, 2, 3$; and $d_{1,2}$ are dipole moments of the transitions $1 \rightarrow 2$, $2 \rightarrow 3$.

We now evaluate a linear response of the system to the field E_2. The equations of motion for the operators a_1, a_2 have the form

$$i \frac{\partial a_1}{\partial t} = -\frac{1}{2} d_1 e_1^* a_2(t) e^{i\Delta t} ,$$

$$i \frac{\partial a_2}{\partial t} = -\frac{1}{2} d_1 e_1 a_1(t) e^{-i\Delta t} , \qquad (5.3.2)$$

$$E_1 = e_1(x, t) \exp(iQx - i\Omega t) ,$$

where

$$\Delta = \Omega = \Omega_1 , \qquad \Omega_1 = \mathcal{E}_1 - \mathcal{E}_2 ,$$

$$a_1(t) = c_1(t) \exp\left(i\mathcal{E}_1 t + \frac{1}{2} iQx \right) ,$$

$$a_2(t) = c_2(t) \exp\left(i\mathcal{E}_2 t - \frac{1}{2} iQx \right) .$$

It will be assumed that E_1, E_2 and the vectors d_1 and d_2 are collinear and that $\Delta = 0$.

Consider the case when the pulse duration τ_p is considerably shorter than the inhomogeneous relaxation time. Then, there exists a soliton mode of pulse propagation (2π-pulses)

$$e_1(\tau) = e_1^*(\tau) = \frac{2}{d_1\tau_p}\mathrm{sech}\frac{\tau}{\tau_p} ,$$

$$n_2(\tau) = \mathrm{sech}^2\frac{\tau}{\tau_p} ,$$

(5.3.3)

where n_2 is the population of the level 2, $\tau = t - x/u$, and u is the pulse velocity. In this case, we may obtain the following relation between the electric induction D_2 and the field strength E_2 in the presence of the soliton:

$$D_2(x,t) = \varepsilon_0 E_2(x,t) + i2\pi d_2^2\frac{N}{V}\mathrm{sech}^2\frac{\tau}{\tau_p}\int_{-\infty}^{t} E_2(x,t)$$

$$\times \exp\left[(i\Omega_2 - \gamma_2)(t_1 - t)\right]dt_1 .$$

(5.3.4)

Here, ε_0 is the nonresonant part of the dielectric permeability, N/V is the atomic density, and γ_2^{-1} is the relaxation time for the transition $3 \rightarrow 2$ and field strength E_2 is necessarily altered by a travelling inhomogeneity.

In the frame of reference moving together with the soliton, frequencies of the incident (ω) and reflected ($\bar{\omega}$) waves are related by the relations:

$$\omega = \omega\frac{1 + \sqrt{\varepsilon_0}u/c}{1 - \sqrt{\varepsilon_0}u/c} ,$$

$$\omega' = \bar{\omega}\frac{1 - \sqrt{\varepsilon_0}u/c}{1 + \sqrt{\varepsilon_0}u/c} , \quad \omega' = \bar{\omega}'$$

to yield

$$\bar{\omega} = \omega\frac{1 + \sqrt{\varepsilon_0}u/c^2}{1 - \sqrt{\varepsilon_0}u/c^2} .$$

(5.3.5)

This formula describes the Doppler effect for the reflection of a weak wave by the soliton.

We define a reflection coefficient for the weak wave scattering from the soliton. If the pulse lengths $\ell_p = u\tau_p$ are much greater than the wavelengths of the weak field $\lambda = 2\pi/k$, the WKB approximation may be applied. Then, an equation for the z-component of the electric field strength E_z takes the form

$$\frac{d^2 E_z}{dx^2} + \left(2i\frac{\omega}{c} - \frac{1}{f}\frac{df}{dx}\right)\frac{dE_z}{dx} + \frac{\omega^2}{c^2}$$

$$\times \left(f^2 - \varphi^2 + i\frac{\omega}{c}f\frac{d}{dx}\frac{\varphi}{f}\right)E_z = 0 ,$$

(5.3.6)

where

$$f(x,\omega) = \frac{1 - u^2/c^2}{1 - \varepsilon u^2/c^2} \ ,$$

$$\varphi = \frac{u(1 - \varepsilon)}{c(1 - \varepsilon u^2/c^2)} \ ,$$

$$\varepsilon(x,\omega) = \varepsilon_0 + \frac{2\pi d^2(N/V)(1 - u^2/c^2)^{1/2}}{\tilde{\Omega} - \omega - uk - i\tilde{\gamma}}$$

$$\times \mathrm{sech}^2\left[\frac{x}{u\tau_p}\left(1 - \frac{u^2}{c^2}\right)^{1/2}\right] \ , \tag{5.3.7}$$

$$\tilde{\Omega} = \Omega\left(1 - \frac{u^2}{c^2}\right)^{1/2} \ , \quad \tilde{\gamma} = \left[1 - \frac{u^2}{c^2}\right]^{1/2} \ .$$

In the WKB approximation the solution of this equation is taken in the form

$$E_z(x) = y(x)f^{1/2}\exp\left[i\frac{\omega}{c}\int\varphi(\xi)d\xi\right] \ . \tag{5.3.8}$$

From this, we derive an equation for $y(x)$

$$\frac{d^2y}{dx^2} + \frac{\omega^2}{c^2}\varepsilon(x,\omega)f^2y = 0 \ .$$

In the range of parameters of $|\tilde{\Omega} - \omega| \gg uk(x)$ this equation is simplified to

$$\frac{d^2y}{dx^2} + \frac{\omega^2}{c^2}\left[\varepsilon_0 + U_0\mathrm{sech}^2\frac{x}{ut_p} \ y\right] = 0 \ , \tag{5.3.9}$$

$$U_0 = \frac{2\pi d^2(N/V)(\Omega - \omega)}{(\Omega - \omega)^2 + \gamma^2} \ ,$$

for which an exact solution is known. Finally, for the reflection coefficient R, we obtain

$$R = \frac{\exp\left\{2\pi(\omega/c)\sqrt{\varepsilon_0}\ell_p(\sqrt{|U_0|/\varepsilon_0} - 1)\right\}}{1 + \exp\left\{(\omega/c)2\pi\sqrt{\varepsilon_0}\ell_p(\sqrt{|U_0|\varepsilon_0} - 1)\right\}} \ . \tag{5.3.10}$$

Here,

$$\Omega - \omega_0 < 0 \ ,$$

$$4|U_0|\left[\frac{\omega}{c}\ell_p\right]^2 \gg 1 \ .$$

For $|U_0|/\varepsilon_0 > 1$ the reflection coefficient is close to unity and for $|U_0|/\varepsilon_0 < 1$ it decreases abruptly to an exponentially small value.

In the case when the three-level system is a low-concentration impurity in a host matrix, a high reflection coefficient is possible only when the low concentration is compensated by the resonance. This demand seems to be rigid and not always realizable in practice. Nevertheless, the idea of utilizing optical solitons as relativistic mirrors is attractive and warrants further study.

5.4 Theory of Superfluorescence

Most solved problems in soliton theory either consider the evolution of an initial "potential" in a nonlinear medium (initial value problem), its subsequent decay into solitons, and the interaction of these solitons with one another, or are concerned with the construction of nonlinear periodic solutions (finite-gap solutions). In the theory of optical solitons, the investigation of the evolution of initially inverted states, their decay into solitons and other nonlinear excitations is of importance. In analyzing such problems, it is necessary to consider a mixed boundary-initial-value problem, but to date, this has received much less attention. In this section, we will study the problem of the evolution of an initial random polarization in an inverted medium. Strictly speaking, this requires a fully quantum description. However, many results can be obtained by using a semiclassical description. The quantum nature of the electromagnetic field is taken into account by a choice of initial population of the medium as a Gaussian function (*Haake et al.*, 1981). Random initial polarization gives rise to the development of an instability whose final stage is the generation of radiation pulses, i.e., amplified spontaneous emission, or loss without mirrors (also called the superfluorescence phenomenon). Consider a pencil-shaped volume which is filled with two-level atoms. This occurs for the case when Fresnel number $F = S/\lambda L \sim 1$. Here, L is the sample length, S is the cross-section, and λ is the wavelength. The analysis of this phenomenon will be carried out using a solution of a mixed boundary-initial-value problem for the Maxwell-Bloch equations.

The Maxwell-Bloch equations are taken in the form

$$\left(\frac{\partial}{\partial t} + \frac{\partial}{\partial x}\right) E = \langle \varrho \rangle , \tag{5.4.1a}$$

$$\frac{\partial \varrho}{\partial t} + i\lambda\varrho = NE ,$$

$$\frac{\partial N}{\partial t} = -\frac{1}{2}\left[E^*\varrho + E\varrho^*\right] ,$$

where the density matrix ϱ for the atomic subsystem is

$$\varrho = \begin{pmatrix} N & \varrho \\ \varrho^* & -N \end{pmatrix} .$$

We now follow *Zakharov et al.* (1984). For this system, a mixed problem is defined by the following initial and boundary conditions:

$$E(t, 0) = E_1(t) ,$$

$$E(0, x) = E_2(x) , \qquad (5.4.1b)$$

$$\varrho(0, x, \lambda) = \varrho_0(x, \lambda) .$$

For the superfluorescence problem, the field incident on the boundary is absent, and the fields $\varrho_0(x, \lambda)$ and $E_2(x)$ are random functions. In this case, $E_2 = 0$, and for $\varrho_0(x, \lambda)$ the following relations are fulfilled:

$$\langle \varrho(x, \lambda) \varrho_0^*(x', \lambda') \rangle = [g(\lambda) N_0]^{-1} \delta(x - x') \delta(\lambda - \lambda') . \qquad (5.4.2)$$

In the case of exact resonance, and with $g(\lambda) = \delta(\lambda)$, one can find a solution of the Maxwell-Bloch equations using the similarity variable substitution

$$E(x, t) = x E(\xi) , \quad N(x, t) = n(\xi) ,$$

$$\varrho(x, t) = \varrho(\xi) , \quad \xi = 2\sqrt{x(t - x)} . \qquad (5.4.3)$$

Substitution of (5.4.3) into (5.4.1) leads to the following set of equations:

$$\xi E' + 2E = 2\varrho , \quad 2\varrho' = \xi n E , \quad 2n' = -\xi \varrho E . \qquad (5.4.4)$$

Let us examine the region of real ξ, since $\text{Im}\{\xi\} \neq 0$ is nonphysical ($t < x$). The set (5.4.4) is now reduced to a second-order equation. To do this, we make the substitution

$$n = \cos \varphi , \quad \varrho = \sin \varphi , \quad E = \frac{2}{\xi} \varphi' , \qquad (5.4.5)$$

so that (5.4.4) becomes

$$\varphi'' + \xi^{-1} \varphi' - \sin \varphi = 0 . \qquad (5.4.6)$$

This is a sine-Gordon equation with nonstationary friction. Let us construct its approximate solutions. The problem reduces to the evolution of an effective particle with unit mass in a periodic potential with damping. The initial randomness in the phase of the polarization becomes a randomness in the initial position of the particle in the periodic potential. Assuming that the initial phase $\varphi_0 \sim N_0^{-1/2}$, we find a maximal point of the first pulse solution $\xi \sim \ln(1/\varphi_0) \gg 1$. Then, we may neglect the second term in (5.4.6) to obtain a solution:

$$\varphi(\xi) = 4 \tan^{-1} \exp(\xi - \xi_0) ,$$

$$E(\xi) = \frac{2}{\xi} \text{sech}(\xi - \xi_0) , \quad \xi_0 \sim \ln \frac{1}{\varphi_0} . \qquad (5.4.7)$$

In the laboratory frame of reference, we see two waves (at the fronts of which $n = 0$; actually, $n = 0$ corresponds to the inverted medium) travel from the point

x_0 in opposite directions. The intensity of the first radiation pulse emitted by the medium is

$$I(t) = 4t_{sf} \left[t \cosh^2 \frac{2\left(t^{1/2} - t_D^{1/2}\right)}{t_{sf}^{1/2}} \right]^{-1} , \tag{5.4.8}$$

where t_D is the time for the delay of a separate pulse to be determined experimentally; $t_{sf} = 8\pi\tau_0/3\lambda^3 Ln_0$ is the superfluorescence time, τ_0 is the spontaneous emission decay time of a single atom.

In case of degenerate transitions the equations are more complicated, for example, if there is a degeneracy of levels which are $j = 2$ multiplets. Then, instead of (5.4.6), we have

$$\varphi'' + \frac{1}{\xi}\varphi'(\xi) = \frac{4}{(4a + b)} \left(a \sin \varphi + \frac{b}{2} \sin \frac{\varphi}{2} \right) ,$$

$$a = \sigma_{-2} + \sigma_2 , \quad b = \sigma_{-1} + \sigma_1 . \tag{5.4.9}$$

Here $\sigma = N_{m\uparrow} - N_{m\downarrow}$ is determined by pumping conditions, $N_{m\uparrow}$ and $N_{m\downarrow}$ are initial populations of the excited and ground states with the angular momentum projection m on the polarization vector direction, respectively. A quantitative analysis of the solutions of (5.4.9) shows that the function E exhibits two maxima. A comparison between the similarity variable solution and the experimental data on the time-dependence of radiation intensity at the sample exit gives a good agreement. Note that these results have been obtained using the assumption that the interaction of waves propagating in opposite directions can be neglected. A theory based on this assumption will accurately describe only the first pulses of radiation.

Superfluorescence theory was developed by *Gabitov* et al. (1985), who showed that the IST could be applied to the Maxwell-Bloch system when the initial and boundary conditions were specified, as in (5.4.1b). They were, therefore, the first to obtain an approximate solution of a mixed problem relying on the IST. This approach was then developed by *Steudel* (1989), who studied the statistics of the pulse delay, and the generated pulse shape. Delay time statistics of superfluorescent pulses has been estimated from the early stage linear dynamics by *Haake* et al. (1981). Consider the set of Maxwell-Bloch equations for which a mixed problem as described by (5.4.6) has been set up. Let a sample have finite dimensions, i.e., $0 \leq x \leq L$, let $t \geq 0$, and assume that $E_1(t) \rightarrow 0$ for $t \Rightarrow \infty$.

The main idea which permits application of the IST to (5.4.1–6) is a reformulation of the conditions for $t < 0$ so as to reproduce the conditions (5.4.1b) for $t = 0$. For this, one needs to solve a mixed "back in time" problem ($t < 0$) by imposing an additional condition $E(t, x) \rightarrow 0$ for $t \rightarrow -\infty$. Note that changing the sign of the time variable also changes the direction characteristics. It is therefore necessary to set up a boundary condition at $x = L$:

$$E_3(t) = E(t, o) , \quad t < 0 ; \quad E_4(t) = E(t, L) . \tag{5.4.10}$$

The latter are chosen so that $E(x, t) \to 0$ is satisfied as $t \to \infty$. From the solution of this problem, one can define the functions $E_3(t)$ and $r(x, \lambda) = \varrho(x, \lambda, t = -\infty) \exp(2i\lambda t)$ to reduce the mixed problem (5.4.1a,b) to the asymptotic one. For the case of small initial conditions a linear approximation can be applied. To construct this solution, let

$$g(\lambda) = \frac{\varepsilon}{\pi(\lambda^2 + \varepsilon^2)} ; \quad E_2(x) = 0 ;$$

$$\varrho_0(x, \lambda) = \varrho_1(\xi) \exp(i\xi x) , \quad |\varrho_1| \ll 1 . \tag{5.4.11}$$

From the solution of linear equations, we have

$$E_3(t) = \frac{\varrho_1}{p_1 - p_2} \exp(-p_2 t) + \theta(-t - L) \exp(-p_1 t) ,$$

$$E_4(t) = \frac{\varrho_1}{p_1 - p_2} \exp(i\xi L) \left[\exp(-p_2 t) - \exp(-p_1 t)(1 - \theta(-t)) \right]$$

$$+ \theta(t) \sqrt{\frac{L}{1 - t}} \frac{I_1(2\sqrt{-Lt})}{2\varepsilon + p_1} \exp(p_1 L + 2\varepsilon t) ,$$

$$p_{1,2} = \frac{i\xi - 2\varepsilon \pm \sqrt{(i\xi + 2\varepsilon)^2 + 4}}{2} , \tag{5.4.12}$$

$$\theta(t) = 0, \ (t < 0) ; \quad \theta(t) = 1, \ (t > 0) .$$

Note the behaviour of the auxiliary pulse $E_3(t)$ incident on the medium. For $-t < L$ this pulse grows exponentially. The requirement for the linear approximation $E_{max} \ll 1$ to be applicable yields

$$|\varrho_0| \exp(L\gamma_{max}) \ll 1 ,$$

$$\gamma_{max} = \max \gamma(\xi) . \tag{5.4.13}$$

With $E_3(t)$, we can – once we have solved a direct scattering problem with the potential $E_0(t) = E_1(t)$, $t > 0$ and $E_0(t) = E_3(t)$, $t < 0$ – define scattering data. Then, the kernel of the Marchenko equation $[F(t, x)]$ is restored, and a value for the field $E(x, t)$ can be derived from the solution of the mixed problem for lasers of moderate length satisfying the condition (5.4.18).

For the case of superfluorescence, we have

$$E_1(t) = 0 , \quad E_2(x) = 0 .$$

The function $\varrho_0(x, \lambda)$ is chosen to model the quantum fluctuations in the laser. It can be taken to be a Gaussian random function with a correlation as in (5.4.2). We can show that

$$F(2t, x) = \frac{1}{4} E(t, x) ,$$

where

$$F(t, x) = \frac{1}{2\pi} \int_{-\infty+i\gamma}^{\infty+i\gamma} d\lambda c(x, \lambda) \, e^{-i\lambda t} \; .$$

Here $c(x, \lambda) = b^*(x, \lambda)/a(x, \lambda)$ represents scattering data for a linear spectral problem

$$\frac{\partial \phi}{\partial t} = i(-I\lambda + H)\phi \; ,$$

where

$$H = \frac{i}{2} \begin{bmatrix} 0 & E \\ -E^* & 0 \end{bmatrix} , \quad I = \begin{bmatrix} 1 & 0 \\ 0 & -1 \end{bmatrix}$$

(for more detailed treatment see *Gabitov* et al., 1986). With $g(\lambda)$ chosen to be a Lorenz function ($T_2^* = 1/\varepsilon$), an expression for $F(x, t)$ for $t > x$ is

$$F(2t, x) = \frac{\pi}{4T_2^*} \int_0^x dx' \int_{-\infty}^{\infty} d\lambda \, \frac{G(t, x - x'; \lambda)}{\lambda^2 + (1/T_2^*)} \, \varrho_0(x, \lambda) \; ; \qquad (5.4.14)$$

$$G(t, x; \lambda) = \theta(t - x) \left\{ I_0(2\sqrt{x(t - x)}) + \left(i\lambda + \frac{1}{T_2^*} \int_0^{t-x} dt' \, I_0(2\sqrt{x(t' - x)}) \right) \right.$$

$$\left. \times \exp\left[\left(i\lambda + \frac{1}{T_2^*} \right)(t - t' - x) \right] \right\} \exp\left(-\frac{t}{T_2^*} \right) \; .$$

As $\varepsilon \to 0$, the kernel grows exponentially for $t \gg x$. In this region the dependence of F on t and x becomes

$$F(2t, x) = x F\left(2\sqrt{x(t - x)} \right) \; ,$$

where

$$F(y) = \frac{\varrho_0(0) I_1(y)}{2y} \; .$$

For $t \leq x$ the kernel F is small in $1/\sqrt{N_0}$; hence in regions where there is considerable pulse energy, its effects may be neglected.

The condition (5.4.13) guarantees the applicability and uniformity of the self-similar approximation.

5.5 Solitons in Stimulated Raman Scattering

The interaction of intense pulses with resonant media reveals a new way of generating short pulses. In many situations, it is possible to generate ultrashort pulses, or solitons and soliton sets.

Consider the case of one-dimensional stimulated Raman scattering (SRS) in an infinite medium (*Chu* and *Scott*, 1975). Let a wave of frequency ω_1 be incident on the medium and let the frequency ω_2 of the scattered wave be less than ω_1, then the frequency of the scattering wave is $\omega_3 = \omega_1 - \omega_2$. Molecules in the medium will be described as harmonic oscillators. The equation of motion for the oscillator is

$$Z_{tt} + \omega_R^2 Z = aE^2 - qE \ . \tag{5.5.1}$$

For the electromagnetic field, we have

$$E_{xx} - \frac{1}{u^2} E_{tt} = 2\mu_0 a(ZE)_{tt} \ . \tag{5.5.2}$$

Here, u is the group velocity of the electromagnetic wave, μ_0 is the magnetic permeability of the medium, and a is a constant.

A solution of (5.5.1,2) is now sought in the form

$$E = \sum_{j=1}^{2} E_j(x,t)e^{i(\omega_j t - k_j x + \gamma_j)} + \text{c.c.} \ , \tag{5.5.3}$$

$$Z = \frac{1}{2} \left[\chi(x,t)e^{i(\omega_3 t - k_3 x + \gamma_3)} + \text{c.c.} \right] \ . \tag{5.5.4}$$

Here, γ_j are constants, and $E_j(x,t)$ and $\chi(x,t)$ are assumed to be gradually varying functions of x and t. Substituting (5.5.3,4) into (5.5.1,2) and keeping only the terms of the first order in the small parameter a, we arrive at the following equations for the field amplitude:

$$\chi_t - i\delta\chi = -iq_3 E_1 E_2^* e^{-i\Delta k x} \ , \tag{5.5.5}$$

$$\frac{1}{u} E_{1t} - E_{1x} = -iq_1 E_2 \chi e^{i\Delta k x} \ , \tag{5.5.6}$$

$$\frac{1}{u} E_{2t} - E_{2x} = -iq_2 E_1 \chi^* e^{-i\Delta k x} \ . \tag{5.5.7}$$

Here,

$$\delta = \frac{(\Delta\omega)^2 - 2(\Delta\omega)\omega_3}{2\omega_3} \ ,$$

$$q_3 = \frac{a}{2\omega_1} \ ; \quad q_j = a\mu_0 \frac{\omega_j - 2\omega_3(\Delta\omega) + (\Delta\omega)}{2k_j} \ ;$$

$$j = 1, 2 ; \quad \Delta\omega = \omega_3 - \omega_R ; \quad k_1 - k_2 - k_3 = \Delta k .$$

Let us assume that

$$\frac{\omega_1}{k_1} = \frac{\omega_2}{k_2} = u , \quad \gamma_1 - \gamma_2 - \gamma_3 = 0 .$$

Introduce the travelling wave variable $\tau = t - x/u$ together with $\zeta = x$, and make the field dimensionless so that

$$A_1 = (q_2 q_3)^{1/2} E_1 e^{-i\Delta k\zeta} ,$$

$$A_2 = (q_1 q_3)^{1/2} E_2 e^{i\Delta k\zeta} , \qquad (5.5.8)$$

$$Y = (q_2 q_1)^{1/2} \chi e^{-i\Delta k\zeta} .$$

We also introduce the bilinear variables

$$U = i A_1 A_2^* , \quad W = A_1 A_1^* - A_2 A_2^* . \qquad (5.5.9)$$

Note that W is the difference in intensities between the incident (or pump) and Stokes waves.

Then, we have the following set of equations:

$$Y_\tau = i\delta Y - U , \qquad (5.5.10)$$

$$U_\zeta = YW - 2i\Delta k U , \qquad (5.5.11)$$

$$W_\zeta = 2(UY^* + U^*Y) . \qquad (5.5.12)$$

These are the basic equations of our model.

Chu and Scott (1975) showed that this system is integrable by the IST. To see this, the set (5.5.10–12) is first written as an operator equation for the Lax pair L and A:

$$iL_t = [L, A] , \qquad (5.5.13)$$

$$L\phi = \lambda\phi , \qquad (5.5.14)$$

$$i\phi_t = A\phi , \qquad (5.5.15)$$

where (5.5.14) has the form

$$\phi_{1\zeta} + i\lambda\phi_1 = Y\phi_2 , \qquad (5.5.16)$$

$$\phi_{2\zeta} - i\lambda\phi_2 = -Y^*\phi_1 ; \qquad (5.5.17)$$

and (5.5.15) is

$$\phi_{1\tau} = \left[\frac{iW}{4(\lambda - \Delta k)} + \frac{i\delta}{2} \right] \phi_1 + \frac{iU\phi_2}{2(\lambda - \Delta k)} ; \qquad (5.5.18)$$

$$\phi_{2\tau} = \frac{iU^*\phi_1}{2(\lambda - \Delta k)} - \left[\frac{iW}{4(\lambda - \Delta k)} + \frac{i\delta}{2} \right] \phi_2 . \qquad (5.5.19)$$

Analysis of the equations for the normalized amplitudes A_1, A_2 shows that they are identical in form to the L-equations given by Lamb for self-induced transparency (*Lamb*, 1980). Indeed, the equations for A_1, A_2 can be written

$$A_{1\zeta} + i\Delta k A_1 = Y(-iA_2) \, ,$$
$$(-iA_2)_\zeta - i\Delta k(-iA_2) = -Y^*(A_1) \, . \tag{5.5.20}$$

This, together with (5.5.17) indicates that these sets of equations coincide if we identify

$$A_1 \equiv \phi_1 \, , \quad -iA_2 \equiv \phi_2 \, , \quad \Delta k \equiv \lambda \, .$$

Once the connection between the two latter SRS equations and the IST L-equation has been established, we can use the IST scheme to find soliton and multi-soliton solutions. Hence, the field $Y(\zeta, \tau)$ can be derived from the Gelfand-Levitan-Marchenko (GLM) equation, taking into account that

$$Y(\zeta, \tau) = -2K(\zeta, \zeta, \tau) \, . \tag{5.5.21}$$

It is required to derive K from

$$K(\zeta, y, \tau) = F(\zeta + y, \tau) - \int_{-\infty}^{\lambda} \int_{-\infty}^{\zeta} F(y + S_1, \tau)$$
$$\times F^*(S_1 + S_2)K(\zeta, S_2, \tau)dS_1 dS_2 \, , \tag{5.5.22}$$

where the kernel F is

$$F(y) = \frac{1}{2\pi} \int_{-\infty}^{\infty} \frac{b(\lambda, \tau)}{a(\lambda, \tau)} e^{-i\lambda y} d\lambda - i \sum_{j=1}^{N} \bar{c}_j(\lambda_j, \tau) e^{-i\lambda_j y} \, .$$

Here, $\bar{c} = b^*(\lambda, \tau)/a_\lambda(\lambda, \tau)$; and N is the number of eigenvalues in (5.5.16–17). The coefficients b and a are defined as follows

$$f = bg + ag^* \, ,$$

where f and g are Jost functions, which satisfy (5.5.17) and have the boundary conditions

$$\lim_{\zeta \to -\infty} f \longrightarrow \begin{pmatrix} 1 \\ 0 \end{pmatrix} e^{-i\lambda\zeta} \, ,$$

$$\lim_{\zeta \to +\infty} g \longrightarrow \begin{pmatrix} 0 \\ 1 \end{pmatrix} e^{i\lambda\zeta} \, .$$

The time-dependence of a, b, and $\bar{c}$ is easily found (from the equation for the A-operator (5.5.16)) to be

$$b(\lambda, \tau) = b(\lambda, 0) \exp\left[\frac{1}{2}i\tau\left(\frac{1}{\lambda - \Delta k} + 2\delta\right)\right] \, , \tag{5.5.23}$$

$$a(\lambda, \tau) = a(\lambda, 0) \,, \tag{5.5.24}$$

$$\bar{c}(\lambda, \tau) = \bar{c}(\lambda, 0) \exp\left(\omega_{Rj}\tau + i\omega_{Ij}\tau\right) \,, \tag{5.5.25}$$

where

$$\omega_{Rj} = \frac{\lambda_{Ij}}{2\left[(\lambda_{Rj} - \Delta k)^2 + \lambda_{Ij}^2\right]} \,,$$

$$\omega_{Ij} = \frac{(\lambda_{Rj} - \Delta k)}{2\left[(\lambda_{Rj} - \Delta k)^2 + \lambda_{Ij}^2\right]} + \delta \,,$$

$$\lambda_j = \lambda_{Rj} + i\lambda_{Ij} \,.$$

The magnitude $b(\lambda, 0)$, $a(\lambda, 0)$, and $\bar{c}(\lambda, 0)$ are defined by the initial conditions.

The single-soliton solution results from setting $b(\lambda, 0) = 0$ and $N = 1$ (5.5.22) and (5.5.21); we derive Y to be

$$Y(\zeta, \tau) = \frac{2i\bar{c}_1(\lambda, 0)}{|\bar{c}_1(\lambda, 0)|} \lambda_{Ij} \exp\left[i(\omega_{I1}\tau - 2\lambda_{R1}\zeta)\right]$$

$$\times \operatorname{sech}\left[2\lambda_{I1}\zeta + \omega_{R1} + \ln\frac{|c_1(\lambda, 0)|}{2\lambda_{I1}}\right] \,.$$

U and W are obtained in a similar way.

5.6 The Evolution of SRS Solitons Under the Action of Molecular Relaxation

From an experimental standpoint, the investigation of the influence of different dissipative mechanisms such as molecular relaxation of SRS soliton dynamics is of much interest. Incorporating damping, a set of equations describing SRS has the form

$$X_\tau = -\varepsilon X + A_1 A_2^* \,, \tag{5.6.1}$$

$$A_{1\zeta} = -X A_2 \,, \tag{5.6.2a}$$

$$A_{2\zeta} = X^* A_1 \,. \tag{5.6.2b}$$

The notation is the same as in the previous section, $X = -iY$, $\Delta k = 0$, $\delta = 0$, and ε is a dimensionless damping coefficient. This parameter may be considered to define the intrinsic time scale, measured in units of the damping time (inverse damping rate). *Druhl* et al. (1983) numerically and experimentally investigated SRS soliton dynamics under the action of damping. Solitons turned out to be spatially broadened under the action of damping, but narrowed in time. Damping of the Stokes wave results in a reinforcement of the pumping wave, and in a generation of solitons.

For small ε, the evolution of SRS solitons under perturbation can be studied analytically using perturbation theory. Since the structure of an associated linear spectral problem is similar to that of the L-operator for the NLS equation, one can develop a corresponding perturbation theory (*Druhl* et al., 1985; *Tadjimuratov* and *Tartakovsky*, 1987; *Kaup*, 1986; *Abdullaev* et al. 1989; see also the articles by *Karpman* and *Maslov*, 1977).

Consider now the stated problem: using linear perturbation theory, we first define the variation of spectral data for the L-operator. Then, solving the GLM equation with modified spectral data, we rebuild the finite wave fields. Let us write out the Jost functions f and g for one-soliton initial conditions:

$$f(x, \lambda) = \frac{e^{i\lambda\zeta}}{\lambda - \lambda_1^*} \begin{pmatrix} \nu e^{i\theta} \operatorname{sech} z \\ \lambda - \mu + i\nu \tanh z \end{pmatrix} , \tag{5.6.3}$$

$$g(x, \lambda) = \frac{e^{-i\lambda\zeta}}{\lambda - \lambda_1^*} \begin{pmatrix} \lambda - \mu - i\nu \tanh z \\ -\nu e^{i\theta} \operatorname{sech} z \end{pmatrix} . \tag{5.6.4}$$

Here, $\lambda_1 = \mu + i\nu$

$$\theta = 2\mu x - \omega_1 \tau - \ln \frac{c(0)}{|c(0)|} + \frac{\pi}{2} , \tag{5.6.5}$$

$$Z = 2\nu x + \omega_2 \tau + \ln \frac{c(0)}{2\nu} .$$

Now we construct an equation for determining the variation of spectral data. Applying a standard procedure (*Karpman* and *Maslov*, 1977), we deduce for the evolution of a single soliton

$$X_s = 2\nu \operatorname{sech} z \exp \left(\frac{i\mu z}{\nu} + i\sigma \right) , \quad z = 2\nu(x - \xi)$$

the equations for the soliton parameters:

$$\mu_\tau = \frac{\varepsilon}{2} \operatorname{Im} \int_{-\infty}^{\infty} dz \, R e^{i\theta} \operatorname{sech} z \tanh z , \tag{5.6.6}$$

$$\nu_\tau = -\frac{\varepsilon}{2} \operatorname{Re} \int_{-\infty}^{\infty} dz \, R e^{i\theta} \operatorname{sech} z , \tag{5.6.7}$$

$$\xi_\tau = -\frac{\omega_R}{2\nu} - \frac{\varepsilon}{4\nu^2} \operatorname{Re} \int_{-\infty}^{\infty} dz \, R e^{i\theta} \operatorname{sech} z , \tag{5.6.8}$$

$$\sigma_\tau = 2\mu\xi_\tau - \omega_I + \frac{\varepsilon}{2} \nu \operatorname{Im} \int_{-\infty}^{\infty} dz \, R e^{i\theta} \operatorname{sech} z (1 - z \tanh z) . \tag{5.6.9}$$

In our case, $\varepsilon R = -\varepsilon X$.

From (5.6.6,7), we derive

$$\mu(\tau) = \mu(0) ,$$
$$\nu(\tau) = \nu(0) \exp(-2\varepsilon\tau) . \tag{5.6.10}$$

The solution of (5.6.8,9) yields

$$
\xi(\tau) = \frac{1}{16\varepsilon\Delta^2} \left(\ln \frac{\Delta^2 + \nu^2(0)}{\Delta^2 + \nu^2(\tau)} - 4\varepsilon\tau \right) ;
$$

$$
\sigma(\tau) = \frac{1}{8\varepsilon\Delta^2} \left(\ln \frac{\Delta^2 + \nu^2(0)}{\Delta^2 + \nu^2(\tau)} - 4\varepsilon\tau \right) + \delta(\tau) ,
$$

(5.6.11)

where $\Delta = \mu - \Delta k$.

As seen from (5.6.11), the soliton amplitude decreases and its spatial width increases. For the time width, we have

$$
\Delta\tau \sim \omega_R = \frac{2(\Delta^2 + \nu^2)}{\nu} = 4\Delta\cosh(2\varepsilon\tau - \eta) .
$$

(5.6.12)

Here, $\eta = \ln(\nu(0)/\Delta)$.

Hence, for $\eta > 0$, $\tau < \eta/2\varepsilon$, and $(\nu(0) > \Delta)$ the soliton narrows in time; this can be used in the generation of solitons in SRS. The following fact is worth noting: *Chu* and *Scott* (1975) reported solutions for SRS solitons as

$$
X = \mu_I \, e^{i\alpha(\zeta,\tau)} \mathrm{sech}\, \beta(\zeta,\tau) ,
$$

$$
A_1 = \mu_I e^{i\alpha(\zeta,\tau)} \mathrm{sech} \left[\frac{\beta(\zeta,\tau)}{\sqrt{\mu_I^2 + \mu_R^2}} \right] ,
$$

(5.6.13)

$$
A_2 = \frac{\mu_I \tanh \beta(\zeta,\tau) - i\mu_R}{\sqrt{\mu_I^2 + \mu_R^2}} .
$$

Here,

$$
\alpha(\zeta,\tau) = \mu_R\zeta + \omega_I\tau , \quad \beta(\zeta,\tau) = \mu_I\zeta + \beta_0 ,
$$

$$
\omega_R = \frac{\mu_I}{\mu_I^2 + \mu_R^2} ; \quad \omega_I = \frac{\mu_R}{\mu_I^2 + \mu_R^2} ;
$$

and μ_R, μ_I are integration constants.

The group velocity is

$$
v = \frac{c}{1 + c(\mu_I^2 + \mu_R^2)} , \quad \Delta\tau = \omega_R^{-1} .
$$

The field values are assumed to be given for all ξ at $\tau = t - x/v = 0$, and their values for $\tau \to 0$ are calculated by the IST. These solutions are determined for characteristics $\tau = 0$ and do not satisfy boundary conditions of the form

$$
A_k(z = 0, t) = A_k(\zeta = 0, \tau = t) = a_k(\tau) , \quad k = 1, 2 ,
$$

(5.6.14a)

$$
a_k(\tau) = 0 , \quad \tau < 0 ,
$$

(5.6.14b)

$$
X(z, t = 0) = 0 , \quad z > 0 ,
$$

(5.6.14c)

$$X(z, t = c/z) = X(\zeta = z, \tau = 0) = 0 . \tag{5.6.14d}$$

Firstly, the solutions (5.6.13) do approximately satisfy (5.6.14) with a fairly good accuracy for a finite range of τ. Secondly, it should be noted that for $\beta_0 \gg 1$, $X(\zeta, \tau = 0)$ will be exponentially small for $\zeta > 0$ and the condition (5.6.14c) is satisfied approximately (*Druhl* et al., 1985) (Fig. 5.1.). These results are obtained in the adiabatic approximation (*Tadjimuratov* and *Tartakovsky*, 1987), where the integrity of the soliton is maintained (see Chap. 2). To determine the variation of the soliton profile, one needs to evaluate a correction to the continuous spectrum (*Abdullaev* et al., 1989).

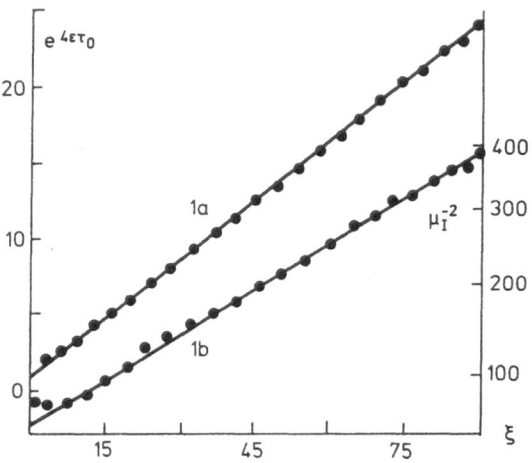

Fig. 5.1. Curve *1a*: soliton position, the solid line is the analytical result, the dots are numerical results; curve *1b*: the analogous results for soliton width $\Delta\tau(\zeta)/\Delta\tau(0)$. From *Druhl* et al., (1985)

A solution is taken to the first order in ε, in the form

$$\begin{aligned} u(x, \tau) &= u_s(x, \tau) + \delta u(x, \tau) \\ &= 2\nu\, e^{i(\mu z/\nu + \sigma)}(\text{sech }z + \omega(z)) , \quad u = iX^* . \end{aligned} \tag{5.6.15}$$

To find the evolution of the continuum spectrum, we need to evaluate the variation of the scattering coefficient b in the Marchenko equation (see Appendix B and Chap. 2). Let us define K and F in (5.5.22) as

$$\begin{aligned} K_1(x, y) &= K_{1s}(x, y) + \delta K(x, y) , \\ F(x) &= F_s(x) + \delta F . \end{aligned} \tag{5.6.16}$$

Here, for single-soliton solution, we have

$$K_{1s} = \frac{2\nu b^*(\lambda_1)e^{-i\lambda_1^*(x+y)}}{1 + |b|^2 e^{-4\nu x}} , \tag{5.6.17}$$

$$F_s = 2\nu b(\lambda_1)e^{i\lambda_1 x} ,$$

$$\delta F = \frac{1}{2\pi} \int_{-\infty}^{\infty} \frac{b(\lambda)}{a(\lambda)} \, e^{i\lambda x} d\lambda \; . \tag{5.6.18}$$

Substituting (5.6.16) into (5.5.22), we obtain an equation for the correction δK in the first order of ε:

$$\delta K_1(x,y) + \int_x^\infty \delta K_1(x,S_1)$$
$$\times \int_x^\infty F_s^*(y+S_1)F_s^*(S_1+S_2)dS_1 dS_2 = \Phi(x,y) \; , \tag{5.6.19}$$

$$\Phi(x,y) = \delta F^*(x,y) - \int_x^\infty K_1(x,S_1)$$
$$\times \int_x^\infty \big[F_s^*(y+S_1)\delta F(S_1+S_2)$$
$$+ \delta F^*(y+S_1)F(S_1+S_2) \big] dS_1 dS_2 \; . \tag{5.6.20}$$

The solution of (5.6.19) is

$$K(x,y) = \Phi(x,y) - b(\lambda_1)e^{i\lambda_1 x}$$
$$\times K_1(x,y) \int_x^\infty \Phi(x,x')e^{i\lambda x'} dx' \; , \tag{5.6.21}$$

Using (5.6.16–18), we arrive at

$$\omega(z) = \frac{\exp\{-i(\sigma + \mu z/\nu)\}}{2\pi i \nu} \int_{-\infty}^{\infty} \frac{b(\lambda)}{a(\lambda)} \left(\frac{\lambda - \mu + i\nu \tanh z}{\lambda - \mu - i\nu} \right)^2$$
$$\times \exp\left\{ i\lambda \left(\frac{S}{\nu} + 2\xi \right) \right\} d\lambda - \frac{\nu \exp\left[i(\sigma + \mu z/\nu) \right]}{2\pi i \cosh^2 z}$$
$$\times \int_{-\infty}^{\infty} \frac{b^*(\lambda)\exp\left[-i\lambda(z/\nu + 2\xi) \right]}{a^*(\lambda)(\lambda - \mu - i\nu)^2} d\lambda \; . \tag{5.6.22}$$

This equation has the same form as the corresponding equation for the NLS equation (*Karpman* and *Maslov*, 1977). this follows from the fact that the $\hat{L}$-operator in this model is analogous to $\hat{L}$ for the NLS equation. The calculation yields the correction $\omega(z)$ to be

$$\omega(z) = -\varepsilon \frac{\Delta^2 + \nu^2}{\nu \cosh z} \left[2z + \frac{i}{\nu} \left(z^2 + \frac{\pi^2}{12} \right) (\Delta + i\nu \tanh z) \right] \; . \tag{5.6.23}$$

Similarly, corrections for the Stokes and pump fields are

$$\delta A_1 = - \frac{i\varepsilon(\Delta + i\nu)}{\nu \cosh z} (\Delta - i\nu \tanh z)$$
$$\times \left(z^2 + \frac{\pi^2}{12} \right) \exp\left\{ i \left(\lambda x - \frac{\mu z}{\nu} - \sigma \right) \right\} \; , \tag{5.6.24}$$

$$\delta A_2 = i\varepsilon(\Delta + i\nu)\left(z^2 + \frac{\pi^2}{12}\right)\exp(i\lambda x) .$$

So far, we have studied SRS soliton dynamics within a simplified model which is not to be formulated in terms of the evolution of the initial data. *Kaup* (1986) attempted to determine the evolution of an arbitrary set of initial data, and to seek the conditions for soliton generation. *Englund* and *Bowden* (1986) studied numerically a process of spontaneous generation of Raman solitons from quantum fluctuations. At present, this branch of physical of optical solitons is being intensively developed, therefore, new results are expected to appear. Unfortunately, experiments considerably lag behind theory; there has been only one paper reporting observation of Raman solitons (*Druhl* et al., 1983).

6. Quantum Optical Solitons

In the previous chapters we applied the IST to find the evolution of solitons and breathers and other localized waves in nonlinear media. In this analysis we did not go beyond the scope of classical physics. However, the IST may also be developed as an exact operational method for solving quantum problems. Many quantum models (NLS, SG, XYZ, etc.), whose classical analogs are integrable systems, have been analyzed. A common programme for the approach and a discussion of the available results on basic quantum models was comprehensively expounded in the reviews of *Faddeev* (1979) and *Thacker* (1981).

Here, in this chapter, the main stages of the development of the quantum IST will be outlined. It should be noted that another approach to exactly solvable models, the so-called Bethe anzatz, was being developed as a method for solving a number of quantum field models and those of quantum statistics.

Bethe formulated this method in 1931, and it was successfully applied by *Lieb* and *Liniger* (1963) to bosons interacting via a δ-shaped pair potential. This problem is equivalent to the solution of the quantum NLS.

Using the quantum IST (QIST), Thirring's massive model, quantum SG, Gross-Neveu model among others, have been studied (*Faddeev*, 1979; *Thacker*, 1981). Faddeev et al. established the relationship between the different approaches using the Bethe anzatz and the QIST. The QIST was shown to be an "algebraization" of the Bethe anzatz.

The next stage in the development of the QIST occurred in the simulation of correlation functions in integrable quantum models such as the one-dimensional boson model, the Heisenberg XXX model, and the antiferromagnet XXZ model (*Korepin*, 1984; *Bogolyubov* and *Korepin*, 1985), these permitting the evaluation of a correlation radius for an arbitrary coupling constant and the determination of the thermodynamic equilibrium state at finite temperature, etc..

This chapter discusses the application of the QIST to quantum nonlinear optics. Before starting with a statement of the QIST, we report the basic principles of the classical IST in a form convenient for quantization. Then, the application of QIST to the NLS equation and the main ideas of this approach are illustrated. Finally, the solutions of the Dicke model and the quantum SG are discussed. This is a reasonable but rather personal restrictions to the topics which are felt to deserve more attention before it is sensible to tackle individual problems, say, like the role played by Raman quantum noise in generating wide-band squeezing for ultrashort pulses as considered by *Carter* and *Drummond* (1991).

6.1 Hamiltonian Formulation of the IST Method and Classical r-Matrix

Here, we present a Hamiltonian approach to the classical IST method *Faddeev*, 1980; *Faddeev* and *Takhtajan*, 1986). The proofs reported in this section are found in the cited literature.

The nonlinear Schrödinger equation in the form

$$iu_t + u_{xx} - 2\kappa|u|^2 u = 0 \tag{6.1.1}$$

is derived from the Hamiltonian

$$H = \int_{-\infty}^{\infty} \left(|u_x|^2 + \kappa|u|^4\right) dx , \tag{6.1.2}$$

and the Poisson brackets (BP)

$$\{u(x), u(y)\} = \{u^*(x), u^*(y)\} = 0 , \quad \{u(x), u^*(y)\} = i\delta(x - y) , \tag{6.1.3}$$

where the PBs are defined by

$$\{F, G\} = i \int_{-\infty}^{+\infty} \left(\frac{\delta F}{\delta u(x)} \frac{\delta G}{\delta u^*(x)} - \frac{\delta G}{\delta U^*(x)} \frac{\delta F}{\delta u(x)}\right) dx . \tag{6.1.4}$$

The PBs satisfy the properties of the antisymmetry and the Jacobi identity:

$$\{F, G\} = -\{G, F\} , \tag{6.1.5}$$

$$\{F\{G, H\}\} + \{H, \{F, G\}\} + \{G, \{H, F\}\} = 0 . \tag{6.1.6}$$

Now, the NLS equation is present as the Hamiltonian motion equation

$$u_t = \{H, u\} = -i\frac{\delta H}{\delta u^*} . \tag{6.1.7}$$

Together with H, we use functionals (see Appendix A):

$$N = \int_{-\infty}^{\infty} |u|^2 dx , \quad P = \int_{-\infty}^{\infty} dx(u_x u^* - u_x^* u) ,$$

which have the meaning of the particle number and momentum. One can see that $\{H, N\} = 0$, $\{H, P\} = 0$. The classical IST is based on the fact that the equation (6.1.1) can be considered as a condition of compatibility of a system of equations

$$F_x = U(x, t, \lambda)F , \tag{6.1.8}$$

$$F_t = V(x, t, \lambda)F , \tag{6.1.9}$$

where $F = \begin{pmatrix} f^{(1)} \\ f^{(2)} \end{pmatrix}$ is the vector function and 2×2 matrices U and V are given by the formulae

$$U = U_0 + \lambda U_1 = \sqrt{\kappa} \begin{pmatrix} o & u^* \\ u & o \end{pmatrix} + \frac{\lambda}{2i} \sigma_3 \; , \tag{6.1.10}$$

$$V = V_0 + \lambda V_1 + \lambda^2 V_2 \; , \tag{6.1.11}$$

$$V_1 = -U_0 \; , \quad V_2 = -U_1 \; ,$$

$$V_0 = i\kappa |u|^2 \sigma_3 - i\sqrt{\kappa}(u_x^* \sigma_+ - u_x \sigma_-) \; ,$$

$$\sigma_\pm = \frac{1}{2}(\sigma_1 \pm i\sigma_2) \; .$$

Here, σ_i is the Pauli matrix

$$\sigma_1 = \begin{pmatrix} 0 & 1 \\ 1 & 0 \end{pmatrix} , \quad \sigma_2 = \begin{pmatrix} 0 & -i \\ i & 0 \end{pmatrix} , \quad \sigma_3 = \begin{pmatrix} 1 & 0 \\ 0 & -1 \end{pmatrix} .$$

The condition of compatibility of a system (6.1.8,9) takes the form

$$U_t - V_x + [U, V] = 0 \; , \tag{6.1.12}$$

and is equivalent to the equation (6.1.1). The condition of (6.1.12) is called a condition of zero curvature. One can show that an equation permitting a representation (6.1.12) possesses an infinite set of the motion integrals. To investigate the equation (6.1.8), let us consider a transition matrix $T(x, y, \lambda)$ at an interval $-L < y < x < L$

$$T(x, y, t) = \exp^{\leftarrow} \int_y^x U(x, t, \lambda) dx \; , \tag{6.1.13}$$

where the sign $\leftarrow$ denotes an order in x, i.e., $U(x)$ moves from the right to the left with an increase of x. The matrix $T_L(\lambda) = T(x = L, y = -L, \lambda)$ is called a monodromy matrix. The matrix U possesses a relationship of involution

$$U(x, \lambda) = \sigma U(x, \lambda^*) \sigma \; ,$$

where $\sigma = \sigma_1$ at $\kappa > 0$; $\sigma = \sigma_2$ at $\kappa < 0$. This property is transferred to the matrix of monodromy; this allows to write it in the form

$$T_L(\lambda) = \begin{pmatrix} a_L & \varepsilon b_L^* \\ \varepsilon b_L & a_L^* \end{pmatrix} , \quad \varepsilon = \text{sign } \kappa \; . \tag{6.1.14}$$

By virtue of the fact that matrix U is traceless, we have $\det T_L = 1$. For decreasing $u(x)$ under real λ there exist some limits

$$b(\lambda) = \lim_{L \to \infty} b_L(\lambda) \; ; \quad a(\lambda) = \lim_{L \to \infty} a_L(\lambda) e^{i\lambda L} \; .$$

The transition matrix $T(x, y, \lambda)$ satisfies the equation

$$T_x(x, y) = U(x) T(x, y) \; , \tag{6.1.15}$$

and the condition $T(x, y)|_{x=y} = 1$, with the obvious properties

$$T(x,z)T(z,y) = T(x,y) \ ,$$

$$T(x,y) = T^{-1}(y,x) \ .$$

Equation (6.1.15) is equivalent to the integral equation

$$T(x,y) = E(x-y) + \int_y^x T(x,z)U_0(z)E(z-y)dz \ , \qquad (6.1.16)$$

$$E(x,\lambda) = \exp\left(\frac{\lambda x}{2i}\ \sigma_3\right) \ ,$$

which can be solved by iterations, and from which one can derive the coefficients $a_L(\lambda)$ and $b_L(\lambda)$ through $u^*(x), u(x)$. A transition from variables u^*, u to $a_L(\lambda), b_L(\lambda)$ is convenient since motion integrals, including the Hamiltonian are expressed through $a_L(\lambda)$, and the Poisson brackets of the coefficients a_L and b_L are calculated explicitly. Knowing the PBs between the elements of T_L matrix, one can easily construct the action-angle variables for Hamiltonian H. There exists a very efficient method to calculate such PBs, based on the classical r-matrix (*Sklyanin*, 1979). Further, it seems to be very convenient to make use of the symbols of tensor matrix product $\otimes$ and, also, the objects of $\{A \overset{\otimes}{,} B\}$ representing a matrix 4×4, consisting of all possible PBs elements of the matrices A and B (Appendix E). Taking into account (6.1.3), one can show that

$$\{U(x,\lambda) \overset{\otimes}{,} U(y,\lambda)\} = i\kappa(\sigma_- \otimes \sigma_+ - \sigma_+ \otimes \sigma_-)\delta(x-y) \ , \qquad (6.1.16)$$

and since

$$\sigma_- \otimes \sigma_+ - \sigma_+ \otimes \sigma_- = \frac{i\kappa}{2(\lambda - \mu)} \ [\hat{P}, \lambda\sigma_3 \otimes I + \mu I \otimes \sigma_3] \ ,$$

where $\hat{P}$ is the permutation operator. Then the relationship (6.1.16) can be rewritten in the form

$$\{U(x,\lambda) \overset{\otimes}{,} U(y,\mu)\} = [r(\lambda - \mu), U(x,\lambda) \otimes I$$
$$+ i \otimes U(y,\mu)]\delta(x-y) \ , \qquad (6.1.17)$$

where

$$r(\lambda) = -\frac{\kappa}{\lambda} \ \hat{P} \ . \qquad (6.1.18)$$

The relationship (6.1.17) is called the fundamental Poisson brackets, which are the universal property of the matrices $U(x,\lambda)$ participating in representation of the zero curvature. Using (6.1.17), one can demonstrate that the PBs of elements of the transition matrix take the form

$$\{T(x,y,\lambda) \overset{\otimes}{,} T(x,y,\mu)\} = [r(\lambda - \mu), T(x,y,\lambda) \otimes T(x,y,\mu)] \ . \qquad (6.1.19)$$

Let us write out some of 16 PBs defined by the expression (6.1.19). While making a limit transition in (6.1.19) $x \to L \to \infty$, $y \to -L \to -\infty$, one obtains

$$\{a(\lambda), a(\mu)\} = 0 ,\tag{6.1.20}$$

$$\{a(\lambda), a^*(\mu)\} = 0 ,\tag{6.1.21}$$

$$\{a(\lambda), b(\mu)\} = \frac{\kappa}{\lambda - \mu + io}\, a(\lambda)b(\mu) ,\tag{6.1.22}$$

$$\{a(\lambda), b^*(\mu)\} = \frac{\kappa}{\lambda - \mu + io}\, a(\lambda)b^*(\mu) ,\tag{6.1.23}$$

$$\{b(\lambda), b(\mu)\} = 0 ,\tag{6.1.24}$$

$$\{b(\lambda), b^*(\mu)\} = 2\pi i |\kappa|\, |a(\lambda)|^2 \delta(\lambda - \mu) ,\tag{6.1.25}$$

and also that

$$\{H, a(\lambda)\} = 0 ,\tag{6.1.26}$$

$$\{H, b(\lambda)\} = i\lambda^2 b(\lambda) .\tag{6.1.27}$$

From (6.1.26,27) and the Hamiltonian equations follow the well known results

$$b(\lambda, t) = \exp\{i\lambda^2 t\} b(\lambda) , \quad a(\lambda, t) = a(\lambda) ,\tag{6.1.28}$$

i.e., $a(\lambda)$ is the generating function of the motion integrals of the dynamical system under consideration. The local integrals of motions are obtained from the expansion coefficients of $\ln a(\lambda)$ in degrees λ^{-1}:

$$\ln a(\lambda) = c_n \lambda^{-n} ,$$

hence, it follows that

$$c_1 = i\kappa N ; \quad c_2 = i\kappa P ; \quad c_3 = i\kappa H .\tag{6.1.29}$$

Note that in this system the action-angle variables are

$$\varrho(\lambda) = \frac{1}{2\pi\kappa} \ln\left(1 + \varepsilon|b|^2\right) , \quad \varphi(\lambda) = -\arg b(\lambda) ,$$

$$\{\varrho(\lambda), \varphi(\mu)\} = \delta(\lambda - \mu) .$$

Thus, while searching for a temporal dependence, we did not use the V-matrix. Moreover, one can show that the existence of the fundamental PB (6.1.17) replaces a condition of the zero curvature, i.e., by the given matrices U and r a set of matrices V_n participating in the zero curvature for all higher NLS equations (i.e., the equations derived from integral motions of the NLS equation) is constructed.

In conclusion, we note that for compatibility of the fundamental Poisson brackets (6.1.17) with properties of antisymmetry and the Jacobi identity (E.12,13) it is enough to perform the relationships

$$r(-\lambda) = -Pr(\lambda)P\tag{6.1.30}$$

and

$$[r_{12}(\lambda - \mu), r_{13}(\lambda) + r_{23}(\mu)] + [r_{13}(\lambda), r_{23}(\mu)] = 0 . \qquad (6.1.31)$$

Equation (6.1.30) is called "classical unitary condition" and (6.1.31) is the "classical Yang-Baxter equation" or "classical triangle equation". One can show that the integrating of a Hamiltonian system corresponds to each solution of the system (6.1.30,31).

6.2 Quantum Nonlinear Schrödinger Equation (QNLSE)

In quantum case the functions $u(x)$ and $u^*(x)$ become Hermite conjugated quantum operators; one obtains the commutational relationship (CR) for them according to the following rule:

$$\{,\} \rightarrow \frac{i}{\hbar}[,] , \qquad (6.2.1)$$

where we further consider that $\hbar = 1$.

The Hamiltonian and elements of the transition matrix (and also its particular case, the monodromy matrix $T_L(\lambda)$) also become the quantum operators, depending upon dynamic variables of a system. In the scheme of the QIST the monodromy matrix also plays a fundamental role, since it helps to construct eigen-functions and spectrum of the Hamiltonian. So, it is of utmost importance to understand the commutational relationship between all the elements of the monodromy matrix. The elements of the matrix $T_L(\lambda)$ are the operators and depend rather complicatedly upon Schrödinger operators u and u^*. Nevertheless, with the help of quantum R-matrix, one explicitly derives commutational relationships for them. The method of R-matrix is applied when CR between elements $T_L(\lambda)$ is presented in the form:

$$R(\lambda, \mu)(T_L(\lambda) \otimes T_L(\mu)) = (T_L(\mu) \otimes T_L(\lambda))R(\lambda, \mu) . \qquad (6.2.2)$$

This relationship generalizes the relationship (6.1.19) for quantum case. In the case of the existence of R-matrix, the existence of the zero curvature representation follows.

Let us formulate a quantum version of the classical IST which can be applied to the NLS equation, following *Thacker* (1981) and *Sklyanin* (1978).

The quantum NLS Hamiltonian takes the form

$$H = \int_{-\infty}^{\infty} \left(u_x^* u_x + \kappa u^*(x)u^*(x)u(x)u(x)\right) dx . \qquad (6.2.3)$$

We start from a normal ordered operator variant of the Zakharov-Shabat equations:

$$\Phi_x =: U\Phi : . \qquad (6.2.4)$$

The components of (6.2.4) take the form

$$\varphi_x^{(1)} = \frac{-i}{2} \lambda \varphi^{(1)} + \sqrt{\kappa} \varphi^{(2)} u ,$$

$$\varphi_x^{(2)} = \frac{i}{2} \lambda \varphi^{(2)} + \sqrt{\kappa} u^* \varphi^{(1)} ,$$

(6.2.5)

where the symbol of normal ordering (: :) indicates that u^* is located on the left, and u on the right, and analogously for

$$\tilde{\Phi} = \begin{pmatrix} \varepsilon \varphi^{(2)} \\ \varphi^{(1)*} \end{pmatrix} , \quad \varepsilon = \text{sign } \kappa .$$

The Jost functions satisfying (6.2.5) are operator functionals of the fields u and u^*. As in the classical case, we require appropriate boundary conditions. Consider an infinite system with a finite number of particles. This corresponds in classical theory to $u(x) \rightarrow 0$ for $x \rightarrow \pm\infty$. Since we deal with operators, this means that only normally ordered products are equal to zero. The asymptotic condition for the operator of Jost functions is

$$\varphi(x, \lambda) \underset{x \rightarrow -\infty}{\longrightarrow} V(x, \lambda) ,$$

(6.2.6)

where

$$\phi = \begin{pmatrix} \varphi^{(1)} & \varepsilon \varphi^{(2)*} \\ \varphi^{(2)} & \varphi^{(1)*} \end{pmatrix} ; \quad V = \begin{pmatrix} e^{-i\lambda x/2} & 0 \\ 0 & e^{i\lambda x/2} \end{pmatrix} .$$

Let us define $G(x, \lambda)$ to be the solution of (6.2.5) satisfying the boundary condition $G(-L, \lambda) = I$, where I is a unit matrix. For $L \rightarrow \infty$

$$\phi(x, \lambda) = G(x, \lambda) V(-L, \lambda) .$$

(6.2.7)

Then, the operators $a(\lambda)$ and $b(\lambda)$ are defined by

$$\tilde{G}(x, \lambda) \underset{\substack{L \rightarrow \infty \\ x \rightarrow \infty}}{\longrightarrow} T(\lambda) = \begin{pmatrix} a(\lambda) & b^*(\lambda) \\ \varepsilon b(\lambda) & a^*(\lambda) \end{pmatrix} ,$$

(6.2.8)

where

$$\tilde{G}(x, \lambda) = V^{-1}(x, \lambda) G(x, \lambda) V(-L, \lambda) .$$

The operator function $\tilde{G}(x, \lambda)$ satisfies a normal ordered integral equation of the form

$$\tilde{G}(x, \lambda) = I + i \int_{-L}^{x} dy : \tilde{U}(y, \lambda) \tilde{G}(y, \lambda) : ,$$

(6.2.9)

where $\tilde{U}$ is defined by

$$\tilde{U} = \begin{pmatrix} 0 & \sqrt{\kappa}u e^{i\lambda x} \\ -\sqrt{\kappa}u^* e^{-i\lambda x} & 0 \end{pmatrix}$$
$$\equiv \sqrt{\kappa}\, e^{-i\lambda x} U(x)\sigma_+ - \sqrt{\kappa}\, e^{i\lambda x} u^*(x)\sigma_- \;. \tag{6.2.10}$$

One can obtain a series of expansions for $\tilde{G}$, a_L and b_L by iteration of (6.2.9). The basic property of the operators a_L and b_L is that they satisfy simple commutation relations between each other, their complex-conjugated values, and the Hamiltonian. To obtain the algebra of commutators for the operators a_L and b_L, consider an operator 4×4 matrix that is a direct product of the Jost functions of (6.2.5) with two different real eigenvalues

$$\lambda = \lambda_1, \lambda_2 \;,$$

$$H_{12}(x) = G(x, \lambda_1) \otimes G(x, \lambda_2) \equiv G_1 \otimes G_2 \;. \tag{6.2.11}$$

The commutation relations can be obtained by comparing with the reverse product G_2 and G_1:

$$H_{21}(x) = G_2 \otimes G_1 \;. \tag{6.2.12}$$

Differentiating (6.2.11) and making use of the fact that G satisfies the Zakharov-Shabat equations (6.2.5), we obtain

$$\frac{\partial}{\partial x} H_{12}(x) =: U_1 G_1 : \otimes G_2 + G_1 \otimes : U_2 G_2 : \;. \tag{6.2.13}$$

An analogous equation for H_{21} results when λ_1 is replaced by λ_2. To achieve the normal ordering it is necessary to evaluate the commutators u and u^* with the Jost functions $G(x, \lambda)$. They are found from the integral equations (6.2.9) and have the form

$$[u(x), G(x, \lambda)] = -\frac{i\sqrt{\kappa}}{2}\, \sigma_- G(x, \lambda) \;, \tag{6.2.14}$$

$$[u^*(x), G(x, \lambda)] = -\frac{i\sqrt{\kappa}}{2}\, \sigma_+ G(x, \lambda) \;. \tag{6.2.15}$$

Taking into account these two relations, one can carry out a normal ordering in (6.2.13) to give

$$\frac{\partial}{\partial x} H_{12} =: \Gamma_{12} H_{12} : \;, \tag{6.2.16}$$

where

$$\Gamma_{12} = U_1 \otimes I + I \otimes U_2 - i\kappa\sigma_+ \otimes \sigma_-$$
$$= i \begin{pmatrix} \frac{\lambda_1+\lambda_2}{2} & \sqrt{\kappa}u & \sqrt{\kappa}u & 0 \\ -\sqrt{\kappa}u^* & \frac{\lambda_1-\lambda_2}{2} & -i\kappa & \sqrt{\kappa}u \\ -\sqrt{\kappa}u^* & 0 & \frac{\lambda_2-\lambda_1}{2} & \sqrt{\kappa}u \\ 0 & -\sqrt{\kappa}\phi^* & -\sqrt{\kappa}\phi^* & -\frac{\lambda_1+\lambda_2}{2} \end{pmatrix} \;. \tag{6.2.17}$$

In a similar way, we find the equation for H_{21}

$$\frac{\partial}{\partial x} H_{21} =: \Gamma_{21} H_{21} : .$$ (6.2.18)

The most important property of the matrix (6.2.17) is that, with the replacement $\lambda_1 \rightarrow \lambda_2$, the relations

$$\Gamma_{21} = R\Gamma_{12}R^{-1} ,$$ (6.2.19)

$$R = \begin{pmatrix} 1 & 0 & 0 & 0 \\ 0 & \beta & \alpha & 0 \\ 0 & \alpha & \beta & 0 \\ 0 & 0 & 0 & 1 \end{pmatrix} ,$$ (6.2.20)

$$\alpha = \frac{\lambda_1 - \lambda_2}{\lambda_1 - \lambda_2 - i\kappa} ,$$ (6.2.21)

$$\beta = \frac{-i\kappa}{\lambda_1 - \lambda_2 - i\kappa} ,$$ (6.2.22)

are satisfied. Equation (6.2.19) is of fundamental importance; it represents a local feature of an exact integrability of the system. From (6.2.16–18) it follows that $RH_{12}(x)R^{-1}$ satisfies the same equation as $H_{21}(x)$. These two magnitudes are equal at $x = -L$ due to the boundary condition and to the definitions (6.2.11) and (6.2.12). Thus, it follows that they are equal everywhere, i.e.,

$$RH_{12}(x) = H_{21}(x)R .$$ (6.2.23)

Consider the limit $L \rightarrow \infty$, $x \rightarrow \infty$ in (6.2.23) so as to obtain an algebraic expression for the commutators of the operators a and b. Within this limit (6.2.23) becomes

$$R_\infty[T(\lambda_1) \otimes T(\lambda_2)] = [T(\lambda_2) \otimes T(\lambda_1)]R_\infty ,$$ (6.2.24)

where $T(\lambda)$ is defined in (6.2.8), and

$$R_\infty = \begin{pmatrix} 1 & 0 & 0 & 0 \\ 0 & 0 & \gamma & 0 \\ 0 & \alpha & 0 & 0 \\ 0 & 0 & 0 & 1 \end{pmatrix} ,$$ (6.2.25)

where

$$\alpha = \frac{\lambda_1 - \lambda_2}{\lambda_1 - \lambda_2 - i\kappa} , \qquad \gamma = \frac{\lambda_1 - \lambda_2 + i\kappa}{\lambda_1 - \lambda_2} .$$

From (6.2.24,25), we derive all commutation relations for the operators a and b:

$$a(\lambda)b(\mu) = \left(1 - \frac{i\kappa}{\lambda - \mu}\right) b(\mu)a(\lambda) ,$$ (6.2.26)

$$a^*(\lambda)b(\mu) = \left(1 + \frac{i\kappa}{\lambda - \mu}\right) b(\mu)a^*(\lambda) , \qquad (6.2.27)$$

$$b^*(\lambda)b(\mu) = \frac{(\lambda - \mu)^2 + \kappa^2}{(\lambda - \mu)^2} b(\mu)b^*(\lambda) , \qquad (6.2.28)$$

$$[a(\lambda)a(\mu)] = [a(\lambda)a^*(\mu)] = [b(\lambda)b(\mu)] = 0 . \qquad (6.2.29)$$

Commutators for a and b with the Hamiltonian are

$$[H, a(\lambda)] = 0 , \qquad (6.2.30)$$

$$[H, b(\lambda)] = \lambda^2 b(\lambda) . \qquad (6.2.31)$$

The representation for the Hamiltonian and other integrals of motion may be obtained by using the expansion

$$\ln a(\lambda) = \sum_{n=1}^{\infty} \frac{c_n}{\lambda^n} , \qquad (6.2.32)$$

and the integral equation (6.2.9). The first three of them take the form

$$N = \frac{1}{i\kappa} C_1 ; \quad P = \frac{1}{i\kappa} C_2 + \frac{1}{2} C_1 ,$$

$$H = \frac{1}{i\kappa} C_3 + C_2 + \frac{i\kappa}{6} C_1 . \qquad (6.2.33)$$

As seen, the relationships (6.2.33) are different from the corresponding classical ones (6.1.29) with coefficients C_n being expressed through previous integrals of motion in the quantum case.

The commutation relations (6.2.30,31) imply that $a(\lambda)$ generates an infinite number of conservation laws and $b(\lambda)$ is a creation operator of the Hamiltonian eigenstates:

$$|\Psi(\lambda_1, \ldots, \lambda_n)\rangle = b(\lambda_1) \ldots b(\lambda_n)|0\rangle . \qquad (6.2.34)$$

The eigenstates of H and N are

$$H|\Psi(\lambda_1 \ldots \lambda_n)\rangle = \left(\sum_{i=1}^{n} \lambda_i^2\right) |\Psi(\lambda_1, \ldots, \lambda_n)\rangle ,$$

$$N|\Psi(\lambda_1 \ldots \lambda_n)\rangle = n|\Psi(\lambda_1 \ldots \lambda_n)\rangle . \qquad (6.2.35)$$

The state (6.2.34) coincides with that obtained using the Bethe ansatz. The operator $a(\lambda)$ is diagonal on the states (6.2.34):

$$a(\lambda)|\Psi(\lambda_1 \ldots \lambda_n)\rangle = \prod_{i=1}^{n} \left(1 - \frac{i\kappa}{\lambda - \lambda_1}\right) |\Psi(\lambda_1 \ldots \lambda_n)\rangle , \qquad (6.2.36)$$

i.e., the wave function satisfies the Schrödinger equation for n particles.

Now, we can construct the ground state for $\kappa < 0$, assuming that λ_i is complex. To obtain the bound state for n particles, it is necessary to take $\Lambda = \sum_i^n \lambda_i$ to be real and to expand λ_i within the distances $i\kappa$ in the imaginary direction, i.e.,

$$\lambda_1 = \frac{\Lambda}{n} + \frac{1}{2}(n-1)i\kappa \,,$$

$$\lambda_2 = \frac{\Lambda}{n} + \frac{1}{2}(n-3)i\kappa \,,$$

$$\vdots$$

$$\lambda_n = \frac{\Lambda}{n} + \frac{1}{2}(n-1)i\kappa \,, \tag{6.2.37}$$

Such a configuration, shown in Fig. 6.1., is referred to as an "n-string". Once symmetrized, the wave function of the n-particle ground state at rest has the form

$$\phi_B(x_1,\ldots,x_n) = \exp\left\{\frac{1}{2}\,\kappa\sum_{i<j<n}|x_1 - x_j|\right\}. \tag{6.2.38}$$

The ground state energy is

$$E = \sum_1^n \lambda_i^2 = \frac{1}{n}\,\Lambda^2 - \frac{n(n^2-1)}{12}\,\kappa^2\,. \tag{6.2.39}$$

In conclusion, we note that the R-matrix should satisfy the Yang-Baxter relations

$$(I \otimes R(\lambda,\mu))(R(\lambda,\nu)\otimes I)(I\otimes R(\mu,\nu)) =$$
$$= (R(\mu,\nu)\otimes I)(I\otimes R(\lambda,\mu))(R(\lambda,\mu)\otimes I)\,, \tag{6.2.40}$$

which are used while calculating the R-matrix in the models of quantum field theory. The relations (6.2.40) are also used in the theory of factorized S-matrix, since it appeared that the matrix of particle-scattering in the quantum field theory is subordinated by the relation (6.2.40) (*Zamolodchikov*, 1979).

6.3 Quantum Sine-Gordon System

It is also convenient to use a Hamiltonian formulation to describe the quantum SG system (*Faddeev*, 1979). We will outline briefly the main results of this approach.

The SG equation results from the Hamiltonian:

$$H = \int_{-\infty}^{\infty}\left(\frac{1}{2}\,\pi^2 + \frac{1}{2}\left(\frac{\partial u}{\partial x}\right)^2 + \frac{m^2}{\beta^2}(1 - \cos\beta u)\right)dx\,. \tag{6.3.1}$$

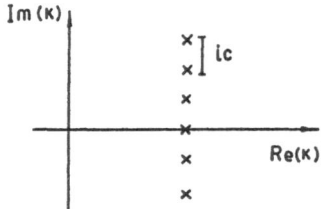

Fig. 6.1. N-string state in the complex λ-plane corresponding to the bound state of N-particle

Here, the parameters m and β stand for the mass and the coupling constant, and

$$\pi = \frac{\partial u(x, t)}{\partial t} \, .$$

Poisson brackets are

$$\{\pi(x), u(y)\} = \delta(x - y) \, . \tag{6.3.2}$$

The linear spectral problem for the SG equation has the form

$$\left[\frac{d}{dx} + \frac{i}{4} \left(\beta \pi \sigma_3 + m \left(\lambda + \frac{1}{\lambda} \right) \right) \sin \frac{\beta u}{2} \, \sigma_1 \right.$$

$$\left. + m \left(\lambda - \frac{1}{\lambda} \right) \cos \frac{\beta u}{2} \, \sigma_2 \right] \Phi = 0 \, , \tag{6.3.3}$$

where σ_1, σ_2, σ_3 are Pauli matrices.

The boundary conditions for (6.3.3) are chosen as

$$u(x) \underset{|x| \to \infty}{\longrightarrow} 0 \left(\mathrm{mod} \, \frac{2\pi}{\beta} \right) \, ; \quad \pi(x) \underset{|x| \to \infty}{\longrightarrow} 0 \, . \tag{6.3.4}$$

The corresponding transition matrix is

$$T(\lambda) = \begin{pmatrix} a(\lambda) & b(\lambda) \\ -b^*(\lambda) & a^*(\lambda) \end{pmatrix} \, , \tag{6.3.5}$$

where $a(\lambda)$ and $b(\lambda)$ are the transition coefficients. For them, the relations

$$a(-\lambda) = a^*(\lambda) \, , \quad b(-\lambda) = -b^*(\lambda) \tag{6.3.6}$$

are valid. The Poisson brackets for $a(\lambda)$ and $b(\lambda)$ are evaluated in the standard way. Thus, we have

$$\{a(\lambda), a(\mu)\} = 0 \, ,$$

$$\{a(\lambda), b(\mu)\} = \frac{\beta^2}{4} \frac{\lambda \mu}{\lambda^2 - \mu^2} \, a(\lambda) b(\mu) \, . \tag{6.3.7}$$

Below, in addition to (6.3.6), it is assumed that the variables

$$\varrho(\lambda) = -\frac{1}{2\pi\gamma\lambda} \ln\left(1 - |b(\lambda)|^2\right) , \tag{6.3.8}$$

$$\phi(\lambda) = -\arg b(\lambda)$$

are canonical. With the expression for $a(\lambda)$, one can find commutational motion integrals. These are evaluated with the aid of the expansion into a series of $\ln a(\lambda)$ for $\lambda \to \infty$. It is required to determine the expansion coefficients:

$$\ln a(\lambda) = \sum C_n \lambda^n ,$$
$$\ln a(\lambda) = \sum C_{-n} \lambda^{-n} . \tag{6.3.9}$$

The Hamiltonian H is

$$H = \frac{2im}{\beta}(C_{-1} - C_1) . \tag{6.3.10}$$

Let us proceed now with the description of the quantum SG approach. The expression for the commutator of the Schrödinger operators $\pi(x)$ and $u(x)$ is written in the form

$$[u(x), \pi(x)] = i\delta(x - y)I , \tag{6.3.11}$$

where I is a unit operator.

The expression for the R-matrix in the SG equation is

$$R = \begin{pmatrix} 1 & 0 & 0 & 0 \\ 0 & b & c & 0 \\ 0 & c & b & 0 \\ 0 & 0 & 0 & 1 \end{pmatrix} . \tag{6.3.12}$$

The R-matrix elements are

$$b = \frac{i\sin\gamma}{\sinh(\alpha - \beta + i\gamma)} , \quad c = \frac{\sinh(\alpha - \beta)}{\sinh(\alpha - \beta + \gamma)} ,$$

$$\alpha = \ln\lambda , \quad \beta = \ln\mu , \quad \gamma = \beta^2 .$$

With this expression one can find the spectrum for the quantum SG system. In particular, the spectrum for the breather energy is given by

$$M_N = \frac{16m}{\gamma} \sin\frac{N\gamma}{16} , \tag{6.3.13}$$

where $N = 1, 2, \ldots, 8\pi/\gamma$. For the case of weak coupling, it follows from (6.3.13) that

$$M_N \approx Nm \left[1 - \frac{1}{6}\left(\frac{N\lambda}{16m^2}\right)^2\right] , \tag{6.3.14}$$

where $\sqrt{\lambda}/m = \beta$.

The quantum SG system describes many physical phenomena. For the case of the self-induced transparency (SIT) in metal vapours, it is possible to obtain the next estimate for the number of breather levels (*Bullough* et al., 1979)

$$N \sim 8\pi\gamma_0^{-1} \sim 2.4 \times 10^7 \,, \quad \gamma_0 \sim 10^{-6} \,,$$

$$M_N = Nmc^2 = N \times 10^9 \text{Hz} \,.$$

This implies that the quantum SG equation describes the propagation of SIT breather pulses, whereas inner oscillations are quantisized, with characteristic distances between the levels of the order of the Lamb shift in the hydrogen atom.

The effect of damping on quantum soliton dynamics in the SG equation has been discussed by *Hida* and *Eckern* (1985).

6.4 The Dicke Model

In this section, using the QIST, we will study a quasi-one-dimensional quantum model of Dicke superradiance in order to show it to be completely integrable. The results will show that, in addition to eigenstates of the system corresponding to states with a continuous spectrum responsible for typical spontaneous radiation, there are also bound quasi-periodic complexes (quantum solitons) which are responsible for the Dicke superradiance pulses.

Dicke superradiance (*Allen* and *Eberley*, 1975; *Andreev* et al., 1978) is the spontaneous emission of a system of two-level atoms interacting via a transverse electromagnetic field. Unlike the usual spontaneous emission of an individual atom, the time of transition of excited atomic ensemble to the ground state is $\tau \propto N_0^{-1}$, where N_0 is the number of atoms, and the superradiance intensity is $\propto N_0^2$.

Prior to the paper by *Rupasov* (1983), several methods for describing superradiance (SR) existed. All the approaches were approximate, and handled one or another scheme of correlator decoupling in a set of Heisenberg motion equations.

Rupasov and *Yudson* (1984), whom we will follow below, showed that a quasi-one-dimensional quantum SR theory belongs to a class of completely integrable models of quantum field theory. Their solution was based on the QIST outlined above.

We now describe the Dicke model in the dipole approximation. A gas of two-level atoms interacting with a transverse electromagnetic field is described by the Hamiltonian

$$H = \frac{\omega_{12}}{2} \int d^3r \left[n(r) + \frac{N_0}{V} \right] + \sum_g \omega_g A_g^+ A_g$$

$$- d \int d^3r \left[E^+(r)P(r) + P^+(r)E(r) \right] \,, \tag{6.4.1}$$

where ω_{12} is the transition frequency and d is the dipole moment of the transition of a two-level atom. The operators $P(r)$ and $n(r)$ satisfy the commutation relations

$$\left[P^+(r), P(r')\right] = n(r)\delta(r - r') ,$$

$$\left[P(r), n(r')\right] = 2P(r)\delta(r - r') , \qquad (6.4.2)$$

$$\left[n(r), P^+(r')\right] = 2P^+(r)\delta(r - r') ;$$

P is the operator of polarization and n the population of a unit volume of the medium, and they are linked to the operators of population n_i and polarization p_i of a single atom by the relations

$$P(r) = \sum_{i=1}^{N_0} p_i\delta(r - r_i) , \quad n(r) = \sum_{i=1}^{N_0} n_i\delta(r - r_i) , \qquad (6.4.3)$$

where N_0 is the total number of atoms of the system. The operator of the electric field strength can be written as an expansion over photon operators:

$$E(r) = i \sum_q \left(\frac{2\pi\omega_q}{V}\right)^{1/2} A_q e^{iqr} , \qquad (6.4.4)$$

where $\omega_q = c|q|$ is the photon frequency and V is the volume of the system. The frequency ω_{12} is assumed to be the same for all atoms in the gas. In (6.4.1), the terms $E^+P^+ + \text{h.c.}$, describing the process of simultaneous creation of a photon with excitation of the atom are omitted. This approximation conforms with the rotating wave approximation. Physically, it is justified by the dominant effect of photons whose frequency is close to that of the transition frequency ω_{12} of the two-level system atom. This allows to restrict ourselves in the expansion of the operator $E(r)$ in (6.4.4) to only the contribution from resonance photons:

$$E(r, t) = i \left(\frac{2\pi\omega_{12}}{V}\right)^{1/2} e^{i(q_0 r - \omega_{12} t)}$$

$$\times \sum_k A_k(t)e^{ikr} = e^{(qr - \omega_{12} t)}\tilde{E}(r, t) , \qquad (6.4.5)$$

where

$$|q_0| = \frac{\omega_{12}}{c} , \quad k = q - q_0 .$$

Expression (6.4.5) is a quantum analog of the transition to slow variables. Note that slow variables $E(r, t)$ and $E^+(r, t)$ are canonical, i.e.,

$$[\tilde{E}(r), \tilde{E}(r')] = 2\pi\omega_{12}\delta(r - r') . \qquad (6.4.6)$$

Separation of a rapidly oscillating part can also be done for the polarization operator $P(r, t)$

$$P(\boldsymbol{r}, t) = p(\boldsymbol{r}, t)e^{i(q_0 r - \omega_{12} t)} .\tag{6.4.7}$$

In (6.4.1) the interaction with only one of the two possible polarizations of the transverse fields has been taken into account. In this section, we will consider the samples with a geometry, such that $F = S/\lambda_0 L >> 1$, where λ_0 is a characteristic radiation wavelenght, and S is a cross-section. All this allows us to consider the interaction of atoms only with those photons whose wave spectrum is aligned along the main axis x of the sample cell. Since the operators $E(x)$ and $E^+(x)$ entering the Hamiltonian (6.4.1) are independent of the transverse coordinates, we can integrate over these, thus permitting the reduction of the problem to a one-dimensional one.

In the following, the system of units $\hbar = c = L/N_0 = 1$ will be used. In the dimensionless variables the Hamiltonian of the Dicke model is

$$H = -i \int_{-\infty}^{\infty} dx \; \varepsilon^+(x)\varepsilon_x(x) - \sqrt{\kappa} \int_{-\infty}^{\infty} dx [\varepsilon^+(x)p(x) + p^+(x)\varepsilon(x)] ,$$

where the interaction constant is $\kappa = 2\pi\omega_{12}d^2/S$, and the dimensionless field operators satisfy the commutation relations

$$[\varepsilon(x), \varepsilon(y)] = \delta(x - y) ; \quad [p(x), n(y)] = 2p(x)\delta(x - y) ,$$

$$[p(x), p(y)] = n(x)\delta(x - y) ;\tag{6.4.8}$$

$$[n(x), n(y)] = 2p^+(x)\delta(x - y) .$$

The first term responsible for the kinetic photon energy leads to a spectral dependence $\omega = k$ in place of the usual spectral dependence $\omega = |k|$. This is not important, since the region of applicability of the model is limited to a parallelogram with sides $\Delta\omega << \omega_{12}$, $\Delta k << q_0 = \omega_{12}$, and centre at the intersection of the spectra of the free fields $\omega = k$, $\omega = \omega_{12}$. In the Hamiltonian (6.4.7), we omit the term

$$\omega_{12} \int_{-\infty}^{\infty} dx \left\{ \varepsilon^+(x)\varepsilon(x) + \frac{1}{2}[n(x) + 1] \right\} \equiv \omega_{12}N ,\tag{6.4.9}$$

which commutes with it. This is equivalent to a change of origin of the energy and momentum, so that the origin of the reference frame k is now at the intersection point of the free field spectra. Then, the Heisenberg equations of motion have the form of the Maxwell-Bloch equations:

$$i(\varepsilon_t + \varepsilon_x) = -\kappa^{1/2}p ,$$

$$ip_t = \kappa^{1/2}\varepsilon n ,\tag{6.4.10}$$

$$in_t = 2\kappa^{1/2}(\varepsilon^+ p - p^+\varepsilon) .$$

A classical analog of this sytem was applied to the analysis of phenomena such as SIT, and photon echo (Chap. 5). In similar problems some emitters are also to be in-phase, i.e., radiation intensity is proportional to N_0^2, but this phasing is due to the external field. Quantum effects have usually been taken into account within a semiclassical approach by adding a random source to (6.4.10). Here, we shall report an exact solution of the operator equations (6.4.10) without using any approximation procedures. Following the method put forward in Sect. 6.1.2, we will find commutator relations for the transition matrix elements of the system.

Let us introduce a two-component field $\Psi_\nu(x)$, $\nu = 1, 2$,

$$\{\Psi_\nu(x), \Psi_{\nu'}(y)\} = \delta_{\nu\nu'} \delta(x - y) \tag{6.4.11}$$

linked with the operators $p(x)$ and $n(x)$ by

$$\begin{aligned} n(x) &= \Psi_2^+(x)\Psi_2(x) - \Psi_1^+(x)\Psi_1(x) \,, \\ p(x) &= \Psi_1^+(x)\Psi_2(x) \,. \end{aligned} \tag{6.4.12}$$

When an additional requirement is imposed on the operators, so that

$$\Psi_1^+(x)\Psi_1(x) + \Psi_2^+(x)\Psi_2(x) = 1 \,, \tag{6.4.13}$$

then the operators N and H in terms of the variables $\varepsilon(x)$ and $\Psi_\nu(x)$ become

$$N = \int_{-\infty}^{\infty} dx [\varepsilon^+(x)\varepsilon(x) + \Psi_2^+\Psi_2] \,, \tag{6.4.14}$$

$$H = -i \int_{-\infty}^{\infty} dx \ \varepsilon^+ \varepsilon_x - \sqrt{\kappa} \int_{-\infty}^{\infty} dx [\Psi_1 \varepsilon^+ \Psi_2 + \Psi_2^+ \varepsilon \Psi_1] \,.$$

Consider a spectral problem for the finite interval $-L \leq x \leq L$:

$$\Phi_x =: U(x, \lambda)\Phi(x, \lambda) : \,, \tag{6.4.15}$$

where the matrix $U(x, \lambda)$ has the form

$$\begin{aligned} U(x, \lambda) = & \ i \begin{pmatrix} 1/2(\lambda - \kappa/\lambda) \,, & \kappa^{1/2}\varepsilon^+ \\ \kappa^{1/2}\varepsilon \,, & -1/2(\lambda - \kappa/\lambda) \end{pmatrix} \\ & + i\frac{\kappa}{\lambda} \begin{pmatrix} \Psi_2^+\Psi_2 & \Psi_2^+\Psi_1 \\ \Psi_1^+\Psi_2 & -\Psi_2^+\Psi_2 \end{pmatrix} \,, \end{aligned} \tag{6.4.16}$$

and the matrix for the solution of (6.4.15) is

$$\Phi(x, \lambda) = \begin{pmatrix} \varphi^{(2)*} & \varphi^{(1)} \\ -\varphi^{(1)*} & \varphi^{(2)} \end{pmatrix} \,. \tag{6.4.17}$$

Now, we will define the matrix $G(x, \lambda)$ to be a solution of (6.4.15) which satisfies the boundary condition $G(x = -L) = I$, here I is a unit matrix. Then, the transition matrix at the finite interval is defined as $G(x, \lambda)$ at the point $x = L$:

$$T_L(\lambda) = G(x = L, \lambda) \,.$$

Based on the results of Sect. 6.2, we can show that the transition matrix elements satisfy the relations

$$R_L(\lambda, \mu)\,(T_L(\lambda \otimes T_L(\mu)) = (T_L(\mu) \otimes T_L(\lambda))\,R_L(\lambda, \mu)\,, \qquad (6.4.18)$$

where

$$R_L = \begin{pmatrix} 1 & 0 & 0 & 0 \\ 0 & \beta & \alpha & 0 \\ 0 & \alpha & \beta & 0 \\ 0 & 0 & 0 & 1 \end{pmatrix},$$

$$\alpha = \frac{\lambda_1 - \lambda_2}{\lambda_1 - \lambda_2 - i\kappa}\,; \quad \beta = \frac{-i\kappa}{\lambda_1 - \lambda_2 - i\kappa}\,.$$

The matrix $R_L(\lambda, \mu)$ is seen to coincide fully with the R_L-matrix of the quantum NLS equation. This coincidence is found also for the infinite interval problem:

$$T(\lambda) = \begin{pmatrix} a^*(\lambda) & b(\lambda) \\ -b^*(\lambda) & a(\lambda) \end{pmatrix}.$$

The corresponding commutation relations retain their shape with $R(\lambda, \mu)$:

$$R(\lambda, \mu) = \begin{pmatrix} 1 & 0 & 0 & 0 \\ 0 & 0 & \alpha & 0 \\ 0 & \gamma & 0 & 0 \\ 0 & 0 & 0 & 1 \end{pmatrix}, \qquad (6.4.19)$$

$$\alpha = \frac{\lambda - \mu}{\lambda - \mu - i\kappa}\,, \quad \gamma = \frac{\lambda - \mu + i\kappa}{\lambda - \mu + i\delta}\,, \quad \delta \to 0\,.$$

In addition, we have

$$[a(\lambda), a(\mu)] = [b(\lambda), b(\mu)] = 0\,,$$

$$[a(\lambda), b(\mu)] = \frac{i\kappa}{\mu - \lambda}\,b(\mu)a(\lambda)\,.$$

It can be shown that the $\ln a(\lambda)$ operator can be expanded in a power series in inverse powers of the spectral parameter, and that the coefficients of the expansion are the commuting integrals of motion. In particular,

$$C_1 = \kappa N\,, \quad C_2 = -i\kappa H\,. \qquad (6.4.20)$$

Now, we proceed with the description of the eigenstates of the Dicke model. The vacuum $|0\rangle$ of the model is a photon-free state where the two-level atoms are all in the ground state:

$$\varepsilon(x)|0\rangle = 0\,, \quad \Psi_2|0\rangle = 0\,, \quad \Psi_1|0\rangle \neq 0\,. \qquad (6.4.21)$$

From the expression for $a(\lambda)$, we have

$$a(\lambda) = 1 + \frac{\kappa}{i\lambda} \int_{-\infty}^{\infty} dx\,\Psi_2^+ \phi^{(2)} e^{-ikx/2}\Psi_2$$

$$+ i\sqrt{\kappa} \int_{-\infty}^{\infty} dx\,\phi^{(1)}\varepsilon e^{-ikx/2} - \frac{\kappa}{i\lambda} \int_{-\infty}^{\infty} dx\,\Psi_1^+ \phi^{(1)}\Psi_2 e^{-ikx/2}\,,$$

where $k = \kappa/\lambda - \lambda$, and we deduce that

$$a(\lambda)|0\rangle = |0\rangle ,$$

(6.4.22)

i.e., $|0\rangle$ is an eigenstate of the a-operator. We introduce the state

$$|\Psi(\mu)\rangle = b(\mu)|0\rangle .$$

Then, using (6.4.20), we obtain

$$a(\lambda, \mu)|\Psi(\mu)\rangle = \frac{\lambda - \mu - i\kappa/2}{\lambda - \mu + i\kappa/2}|\Psi(\mu)\rangle ,$$

(6.4.23)

i.e., $\Psi(\mu)$ is also an eigenstate of the a-operator. Next, we have the relations

$$H|\Psi(\mu)\rangle = -\mu|\Psi\rangle , \quad N|\Psi\rangle = |\Psi\rangle .$$

This indicates that the spectral parameter μ corresponds to the energy, and $\Psi(\mu)$ is a one-particle state. A spectrum of the one-particle state consists of two polarization branches:

$$\omega_{1,2}(k) = \frac{k}{2} \pm \left[\left(\frac{k}{2}\right)^2 + \kappa\right]^{1/2} .$$

(6.4.24)

A multi-particle state is built up analogously:

$$|\Psi(\mu_1, \ldots, \mu_n)\rangle = b(\mu_1)b(\mu_2)\ldots b(\mu_n)|0\rangle .$$

(6.4.25)

As in the quantum NLS model, quasi-particles can form a bound m particle state $|\Psi(\mu)\rangle$ in which μ_ℓ is a complex magnitude,

$$\mu_\ell = \frac{\mu}{m} + i\kappa\left(\frac{m+1}{2} - \ell\right) , \quad \mu = \sum_\ell^m \mu_\ell ,$$

(6.4.26)

and power ω is related to a pulse per particle $k = 1/m \sum k_\ell$ by the expression

$$k = \omega\left\{1 - \frac{\kappa}{m}\sum_{\ell=1}^m \left[\omega^2 + \kappa^2\left(\frac{m+1}{2} - \ell\right)^2\right]^{-1}\right\} .$$

(6.4.27)

The bound state (6.4.25) may be referred to as a quantum soliton. For $m \gg 1$, $\omega \simeq k$ the wavefunction of the bound state has a spatial dimension

$$r_0 \propto (\kappa m)^{-1} .$$

Following the analogy with the results of the classical IST, we believe that an almost completely inverted initial state of the system is likely to disintegrate into a set of quantum solitons with a small admixture of continuous spectrum states. The SR pulses ought to contain a major portion of the energy accumulated in the atomic subsystem.

In a similar way, one can prove complete integrability of the quantum model of superradiance with discrete atoms (*Rupasov* and *Yudson*, 1984).

7. Conclusion

The present monograph reviews recent investigations of optical solitons in non-resonant media and fibres, as well as a number of related problems concerning the effects of resonance impurities in fibres, autowave processes, and so on. The main focus is on problems of propagation, interaction, and generation of solitons and their sets in optical fibres.

The basic features of soliton processes, which occur in nonlinear optical fibres, have been thoroughly studied. A theory of optical solitons in single-mode fibres based on a unified equation – a generalized NLS equation which governs the evolution of a slow electric field envelope – and which describes a wide class of picosecond and subpicosecond processes has been presented. We discuss soliton propagation in fibres, taking into account dissipation, higher order dispersion, nonuniformities, etc. The general approach is also effective in the investigation of the interaction of solitons, and in the study of their generation in fibres from initial complex field distributions. Experiments with optical solitons agree well with theoretical results, which testifies to the efficency of the method.

The investigation of other aspects raised in the book is important in future developments of the subject. In particular, the dynamics of multi-soliton sets and nonlinear periodic waves in optical fibres are important fields of study. Here, theory and experiment are both still in progress.

Generators of optical solitons (the soliton laser) which allow the generation of picosecond and subpicosecond pulses of desired duration and frequency have now been developed. The use of soliton mechanisms enables one to obtain SP compression of the order of 20 fs, and pulses of intensities of $\sim 10^{12}$ W/cm^2. Lasers which generate stable SPs are quite promising for possible applications to atomic and molecular spectroscopy, and fast processes in condensed media.

The soliton laser is one of the first experimental realizations of optical solitons in resonators. The investigation of optical solitons and nonlinear periodic waves in resonators is still an open problem, which should see new significant results in the near future.

Energy pumping into optical solitons to compensate for loss during soliton propagation over long distances has been considered theoretically and, to some extent, solved experimentally. Use is made of energy pumping by the SRS mechanisms, where the energy accumulated in the molecular oscillations of a medium is transferred to an optical soliton. Using an optical fibre, an SRS-laser has now been developed. A similar result is obtained when resonant impurities are incorporated into the optical fibre. The accumulated energy is transferred from these impurities to an USP during its propagation in the fibre, thus fully compensating the dissipation loss.

Note also that in active media the character of the soliton interaction is fundamentally changed – it ceases to be elastic, a continuous spectrum is generated, and stable multi-solitons are formed. These problems in nonlinear optics have still not been studied. In this connection, a recently developed nonlinear theory of superfluoresence is of importance.

Further investigations of dynamic chaos of nonlinear electromagnetic waves, both in a resonator and in an unbounded modulated medium are of much interest.

The IST allowed the solution of a number of problems in quantum nonlinear optics (superfluorescence, self-induced transparency). In our opinion, a promising aspect is the search for new integrable quantum optical systems, as well as the development of approximate methods that could describe quantum solitons in near-integrable systems. The development of an appropriate perturbation theory would allow the solution of many problems, e.g. the various extensions of the Dicke model and chaos in quantum systems.

Finally, it should be pointed out that many problems closely associated with those studied in this book remain beyond the scope of our considerations. Among these are: methods of USP compression in fibres, control of their spectrum and envelope, and the evolution of parametric solitons. The investigation of soliton dynamics in coupled fibres is particularly interesting because of its application to optical processing devices. A linear formulation of the problem of wave propagation in fibres with random parameters raises a number of new effects (an optical analog of Anderson's localization of waves (*Abdullaev* and *Abdullaev* 1980). A nonlinear case will require the solution of problems for multi-dimensional optical solitons.

The next important problem is the investigation of the interaction between optical solitons and other kinds of excitations (phonons, excitons, etc.). *Burlak* et al. (1986) showed that acousto-electromagnetic solitons might exist in fibres. These problems become attractive and will certainly be much studied in the near future.

Appendix

A. Optical Fibres

There is already much in the literature (*Marcuse*, 1972; *Unger*, 1977; *Adams*; 1981; *Snyder* and *Love*, 1984) on optical fibres. There, the properties of planar and circular fibres, as well as their various fields of application have been extensively discussed.

The simplest optical fibre is a circular dielectric cylinder, called a core, placed into a dielectric cladding. For waveguide light propagation it is required that the refraction index of the core n_1 be higher than that of the cladding n_2, so as to give total internal reflection. There are two distinctive types of fibre: the step index fibre, where the core index n_1 is independent on the transverse coordinate, and the graded index fibre whose $n_1 = n_1(p)$ changes continuously from the centre to the core edge.

Consider the step-index fibre. The core radius is denoted by ϱ, and the axis of the coordinate z is aligned with the optical axis. The solution of the wave equation

$$\Delta E_z - \frac{1}{c^2} \frac{\partial^2 D_z}{\partial t^2} = 0 \tag{A.1}$$

for the z-component of the electric vector is sought in the form

$$E_z = A R(\varrho, \varphi) e^{i(qz - \omega t)} , \tag{A.2}$$

where z, ϱ and φ are cylindrical coordinates. Setting $R(\varrho, \varphi) = F(\varrho) e^{i\nu\varphi}$, from (A.1,2) it follows that $F(\varrho)$ obeys the Bessel equation

$$\frac{d^2 F}{d\varrho^2} + \frac{1}{\varrho} \frac{dF}{d\varrho} + \left(\kappa_\ell^2 - \frac{\nu^2}{\varrho^2} \right) F = 0 , \tag{A.3}$$

where

$$\kappa_\ell^2 = k_\ell^2 - q^2 , \quad k_\ell^2 = \frac{\varepsilon_i \omega^2}{c^2} , \quad i = 1, 2 .$$

An analogous equation results for

$$H_z = B R(\varrho, \varphi) e^{i(qz - \omega t)} .$$

Assuming that the solution (A.3) inside the core is the Bessel function $I_\nu(\kappa\varrho)$ and, in the region of cladding it is a Hankel function of the first kind $H^{(1)}(i\gamma\varrho)$ with the imaginary argument $\kappa = i\gamma$, and that these satisfy the boundary conditions for $\varrho = a$, we obtain a set of equations for the amplitudes of the fields E and H in the core and cladding. The condition of solvability of this set of equations gives an expression for eigenvalues that are representative of fibre modes. The modes of an optical fibre are known to have all six field components that cannot be brought into transverse electric (TE) or transverse magnetic (TM) modes, except for the case when $\nu = 0$. Each mode is characterized by a cut-off frequency, i.e., the frequency below which the mode ceases to exist. The peculiarity of an optical fibre is the presence of the HE_{11} mode, whose cut-off frequency is zero. Hence, for frequencies ranging from zero to the cut-off of the next mode, only a single mode can propagate. This is referred to as a single-mode regime. The range of frequencies where the single-mode regime is realized is defined as

$$0 < f < \frac{2.405}{2\pi a\sqrt{\varepsilon_1 - \varepsilon_2}} \,. \tag{A.4}$$

As seen from (A.4), the smaller the difference between dielectric core and cladding constants, and the smaller the core radius, the larger the frequency region of the single-mode regime. It is important to increase the upper limit of this region, since the greater the mode frequency is compared to the cut-off frequency (which is zero for the HE_{11} mode), the greater the portion of the mode field concentrated inside the core.

The structure of the HE_{11} field mode is complicated. However, for the fibres with $\varepsilon_1 - \varepsilon_2 << \varepsilon_1$, one can, to good approximation, assume that the mode is polarized in the plane which is transverse to the fibre axis, and that it has a propagation constant $g \approx k$. Thus, the HE mode can be treated as a quasi-transverse electromagnetic wave (TEM), and in the simplest case is polarized only in either of the transverse directions. In calculations for an approximate description of the transverse distribution of the HE mode, it is convenient to use a Gaussian form

$$R(\varrho) \approx \exp\left(-\frac{\varrho^2}{2\varrho_0^2}\right) \,,$$

where ϱ_0 is an effective radius (*Snyder* and *Love*, 1984).

An important feature of any fibre is its energy loss. There are many causes of energy loss in fibres. These include: impurity absorption and self-absorption by the fibre material, Rayleigh scattering by refractive index fluctuations, and scattering due to different variations in the fibre geometry (variations in the transverse cross-section radius, microbendings, etc), as well as various nonlinear scattering processes occurring at high transmission powers. It is, of course, not possible to eliminate all of these, however, progress in fibre technology permits fabrication of fibres with losses of 0.2 dB/km at the wavelength $\lambda \sim 1.5$ μm. The fabrication of low-loss optical fibres allows many possibilities of utilizing them in optical communication systems, as well as to take account of the loss using perturbation theory to treat it as a small correction when evaluating propagation characteristics in such fibres.

B. The Nonlinear Schrödinger (NLS) Equation

In this appendix we supply background material on the NLS equation required for understanding this work. The NLS equation has the form

$$iq_t + q_{xx} + 2|q|^2 q = 0 . \tag{B.1}$$

Let us suppose that

$$\int_{-\infty}^{\infty} |q_0(x,0)| dx < \infty .$$

The solution of the initial-value problem for rapidly vanishing initial conditions is reduced to the investigation of a spectrum for a set of two first-order differential equations (Zakharov-Shabat system):

$$\phi_x^{(1)} = i\lambda\phi^{(1)} + iq_0(x)\phi^{(2)} ,$$
$$\phi_x^{(2)} = -i\lambda\phi^{(2)} + iq_0^*(x)\phi^{(1)} . \tag{B.2}$$

Discrete values λ_n are representative of solitons and N-soliton states formed during the evolution of a rapidly vanishing initial condition; continuous values of λ are responsible for a continuous spectrum of waves excited in the system. We will specify the most characteristic nonlinear modes used in this book.

1. One-soliton state: The relevant spectral parameter is $\lambda_1 = \mu + i\eta$. The one-soliton solution has the form

$$q_s(x,t) = 2\eta \operatorname{sech} 2\eta(x - x_0 + 4\mu t) \exp\{-2i\mu x - 4i(\mu^2 - \eta^2)t - i\delta_0\} . \tag{B.3}$$

Here, η is the soliton amplitude, $v = -4\mu$ is its velocity.

2. Two-soliton state: The solution can be written

$$q(t, x) = 4\exp\left(-\frac{it}{2}\right) [\cosh(4x) + 4\cosh(2x)$$
$$+ 3\cos(4t)]^{-1}[\cosh(3x) + 3\exp(-4it)\cosh(x)] . \tag{B.4}$$

The modulus of this solution is periodical in t with period $T = \pi/2$.

3. The case of large, smooth initial potential: Let the initial potential satisfy the condition $q_x/q \ll \lambda$. Then, the η_n values are defined by the Bohr quantization rule

$$\int_{-\infty}^{\infty} \sqrt{|q_0|^2 + \eta_n^2} \, dx = 2\pi\left(n + \frac{1}{2}\right) . \tag{B.5}$$

The total number of levels which coincides with the soliton number is

$$N = \frac{1}{\pi} \int_{-\infty}^{\infty} |q_0(x)|\, dx \ . \tag{B.6}$$

The condition for the absence of solitons is

$$\int_{-\infty}^{\infty} |q_0(x)|\, dx < \ln(2 + \sqrt{3}) \ . \tag{B.7}$$

There exists an infinite number of integral invariants of the NLS equation. The first three correspond to the "number of quanta" N, the field momentum P, and the total Hamiltonian H:

$$N = \int_{-\infty}^{\infty} |q|^2 dx \ , \tag{B.8}$$

$$P = \frac{i}{2} \int_{-\infty}^{\infty} (q^* q_x - q_x^* q)\, dx \ , \tag{B.9}$$

$$H = \int_{-\infty}^{\infty} \left(|q_x|^2 - |q|^4 \right)\, dx \ . \tag{B.10}$$

For the case of a single soliton, substitution of (B.3) into (B.8–10) gives

$$N = 4\eta \ , \quad P = 8\mu\eta \ , \quad H = 32\eta \left(\frac{1}{3}\eta^2 - \mu \right) \ . \tag{B.11}$$

The Jost functions for the single soliton initial state have the form

$$\Psi = \begin{pmatrix} \Psi^{(1)} \\ \Psi^{(2)} \end{pmatrix} = \frac{e^{i\lambda x}}{\lambda - \mu + i\eta} \begin{pmatrix} \lambda - \mu + i\eta \tanh 2\eta(x - x_0) \\ \eta \operatorname{sech} 2\eta(x - x_0) e^{-2i\mu x - i\delta_0} \end{pmatrix} \ , \tag{B.12}$$

$$\phi(x, \lambda) = a(\lambda)\tilde{\Psi}(x, \lambda) \ , \tag{B.13}$$

$$a(\lambda) = \frac{\lambda - \mu - i\eta}{\lambda - \mu + i\eta} \ , \tag{B.14}$$

$$\tilde{\Psi} = \begin{pmatrix} -\Psi^{(2)*} \\ \Psi^{(1)*} \end{pmatrix} \ .$$

The "action-angle" variables are

$$n(\lambda) = \frac{1}{\pi} \ln \frac{1}{|a(\lambda)|^2} \ , \tag{B.15}$$

$$\phi(\lambda) = \arg b(\lambda) \ , \tag{B.16}$$

$$N_k = 2\lambda_k \ ; \quad \Phi_k = \ln \frac{1}{b_k} \ , \quad k = 1, 2, \ldots, N \ .$$

Introducing new variables:

$$\bar{N}_k = i(N_k^* - N_k) \ ; \quad \mu_k = \frac{\Phi_k^* - \Phi_k}{2i} \ , \tag{B.17}$$

we find integrals of the motion

$$N = \sum_k \bar{N}_k + \int_{-\infty}^{\infty} n(\lambda)d\lambda \,, \tag{B.18}$$

$$P = i \int_{-\infty}^{\infty} q q_x^* dx = \sum_k \frac{v_k \bar{N}_k}{2} - 2 \int_{-\infty}^{\infty} \lambda n(\lambda)d\lambda \,,$$

$$E = H = \sum_k \left(-\frac{\bar{N}_k^3}{12} + \frac{\bar{N}_k v_k^2}{4} \right) + 4 \int_{-\infty}^{\infty} \lambda^2 n(\lambda)d\lambda \,,$$

where $v_k - 4\text{Re}\{\lambda_k\}$ is the velocity of the k-th soliton, $N_k/2$ is the soliton mass, $\bar{N}_k v_k^2/4$ is its kinetic energy and $\bar{N}_k^3/12$ is its rest energy. The magnitude $[n(\lambda/2)]d\lambda$ is interpreted as the mass of particles moving with velocities $(-4\lambda, -4(\lambda + d\lambda))$. In nonlinear optics normal dispersion implies $d^2\omega/dk^2 > 0$. The NLS equation then has the opposite sign in the nonlinear term

$$iq_t + q_{xx} - 2|q|^2 q = 0 \,. \tag{B.19}$$

This equation has the "dark" soliton solution

$$\begin{aligned} q_s(x,t) &= 2\eta \tanh[2\eta(x - x_0 + 4\mu t)] \\ &\times \exp\{-2i\mu x - 4i(\mu^2 + 2\eta^2)t - i\delta_0\} \,. \end{aligned} \tag{B.20}$$

Consider next the formulae describing the evolution of the NLS solitons under weak perturbations

$$iq_t + q_{xx} + 2|q|^2 q = i\varepsilon R(q) \,; \quad \varepsilon \ll 1 \,. \tag{B.21}$$

The one-soliton solution is taken in the form

$$q_s(x,t) = 2\eta(t)\text{sech}[2\eta(x - \zeta(t))] \exp \left\{ \frac{iz\mu(t)}{\eta(t)} + i\delta(t) \right\} \,, \tag{B.22}$$

where $z = 2\eta(x - \zeta)$. The application of perturbation theory to solitons based on the IST (*Karpman* and *Maslov*, 1977; *Kaup* and *Newell*, 1978) and in the adiabatic limit permits the following equations for the soliton parameters:

$$\frac{d\mu}{dt} = \frac{1}{2} \varepsilon \text{Im} \left\{ \int_{-\infty}^{\infty} dz\, R e^{-i\theta} \text{sech}\, z \tanh z \right\} \,, \tag{B.23}$$

$$\frac{d\eta}{dt} = \frac{1}{2} \varepsilon \text{Re} \left\{ \int_{-\infty}^{\infty} dz\, R e^{-i\theta} \text{sech}\, z \right\} \,, \tag{B.24}$$

$$\frac{d\zeta}{dt} = 4\mu + \frac{\varepsilon}{4\eta^2}\text{Re} \left\{ \int_{-\infty}^{\infty} dz z\, \text{sech}\, z\, R e^{-i\theta} \right\} \,, \tag{B.25}$$

$$\frac{d\delta}{dt} = 2\mu \frac{d\zeta}{dt} - 4(\mu^2 - \eta^2)$$

$$+ \frac{\varepsilon}{2\eta} \operatorname{Im} \left\{ \int_{-\infty}^{\infty} dz \ \operatorname{sech} z(1 - z \tanh z) Re^{-i\theta} \right\} ; \qquad \text{(B.26)}$$

$$\theta = \frac{\mu}{\eta} z + \delta .$$

Let us also state the equations of perturbation theory for the evolution of nonlinear wave packets in a soliton-free region (*Malomed*, 1986). The perturbed evolution equations for the "action-angle" variables $n(\lambda)$ and $\varphi(\lambda)$ are

$$\frac{dn}{dt} = \frac{2}{\pi} \int_{-\infty}^{\infty} dx \ \operatorname{Im} \left\{ \frac{b(\lambda)}{a(\lambda)} [\varepsilon(\phi_2^*(x,\lambda))^2 R(x) \right.$$

$$\left. - \varepsilon^*(\phi_1^*(x,\lambda))^2 R^*(x)] \right\} , \qquad \text{(B.27)}$$

$$\frac{d\varphi}{dt} = 4\lambda^2 + \operatorname{Re} \left\{ \int_{-\infty}^{\infty} dx \frac{a(\lambda)}{b(\lambda)} [\varepsilon^*(\phi_2^*(x,\lambda))^2 R^*(x) \right.$$

$$\left. - \varepsilon(\phi_1(x,\lambda))^2 R(x)] - 2\varepsilon\phi_1(x,\lambda)\phi_2^*(x,\lambda)R(x) \right\} . \qquad \text{(B.28)}$$

Following the special cases of initial conditions, $n(\lambda)$ is either

$$q_0(x) = \begin{cases} a_0 , & 0 \le x \le \ell_0 , \\ 0 , & x < 0 \quad x > \ell_0 , \end{cases} \qquad \text{(B.29)}$$

$$n(\lambda) = \frac{1}{\pi} \left\{ 1 + \left[\sin^2 \left(\ell_0 \sqrt{\lambda^2 + a_0^2} \right) \right] \right.$$

$$\left. \times \left[\frac{\lambda^2}{a_0^2} + \cos^2 \left(\sqrt{\lambda^2 + a_0^2} \right) \right]^{-1} \right\} ;$$

or

$$q_0(x) = a_0 \operatorname{sech} \frac{x}{\ell_0} \exp(igx) ,$$

$$n(\lambda) = -\frac{1}{\pi} \ln \left[1 - \operatorname{sech}^2 \left(\pi\ell_0\lambda - \frac{g}{2} \right) \sin^2(\pi a_0 \ell_0) \right] . \qquad \text{(B.30)}$$

Finally, we report expressions for the variational derivatives written in terms of Jost coefficients and the spectral parameter λ, and also the second variational derivatives of the coefficient $a(\lambda)$. These are all used in this book:

$$\frac{\delta a(\lambda)}{\delta q(x)} = i\phi_1(x)\varphi_1(x) , \qquad \frac{\delta a(\lambda)}{\delta q^*(x)} = -i\phi_2(x)\varphi_2(x) ,$$

$$\frac{\delta b(\lambda)}{\delta q(x)} = i\phi_2^*(x)\varphi_1(x) , \qquad \frac{\delta b(\lambda)}{\delta q^*(x)} = i\phi_1^*(x)\varphi_2(x) , \qquad \text{(B.31)}$$

$$\frac{\delta^2 a(\lambda)}{\delta q(x_1)\delta q(x_2)} = a(\lambda)\big[-\phi_1(x_2)\phi_1(x_1)\tilde{\phi}_1(x_1)\tilde{\phi}_1(x_2)$$
$$+ \theta(x_1 - x_2)\phi_1^2(x_1)\tilde{\phi}_1^2(x_2) + \theta(x_2 - x_1)\phi_1^2(x_2)\tilde{\phi}_1^2(x_2)\big]$$
$$+ b(\lambda)\left[\theta(x_1 - x_2) - \theta(x_2 - x_1)\right]$$
$$\times \left[\phi_1(x_1)\tilde{\phi}_1(x_2) - \phi_1(x_2)\tilde{\phi}_1(x_1)\right]\phi_1(x_2)\phi_1(x_1) , \qquad \text{(B.32)}$$

$$\frac{\delta^2 a(\lambda)}{\delta q^*(x_1)\delta q^*(x_2)} = a(\lambda)\big[-\phi_2(x_1)\phi_2(x_2)\tilde{\phi}_2(x_1)\tilde{\phi}_2(x_2)$$
$$+ \theta(x_1 - x_2)\phi_2^2(x_1)\tilde{\phi}_2^2(x_2) + \theta(x_2 - x_1)\phi_2^2(x_2)\tilde{\phi}_2^2(x_1)\big]$$
$$+ b(\lambda)\left[\theta(x_2 - x_1) - \theta(x_1 - x_2)\right]$$
$$\times \left[\phi_2(x_2)\tilde{\phi}_2(x_1) - \phi_2(x_1)\tilde{\phi}_2(x_2)\right]\phi_2(x_1)\phi_2(x_2) , \qquad \text{(B.33)}$$

$$\frac{\delta^2 a(\lambda)}{\delta q^*(x_1)\delta q(x_2)} = a(\lambda)\big[\phi_1(x_2)\phi_2(x_1)\tilde{\phi}_1(x_2)\tilde{\phi}_2(x_1)$$
$$- \theta(x_1 - x_2)\phi_2^2(x_1)\tilde{\phi}_1^2(x_2) - \theta(x_2 - x_1)\phi_1^2(x_2)\tilde{\phi}_2^2(x_1)\big]$$
$$+ b(\lambda)\left[\theta(x_2 - x_1) - \theta(x_1 - x_2)\right]$$
$$\times \left[\tilde{\phi}_1(x_2)\phi_2(x_1) - \phi_1(x_2)\tilde{\phi}_2(x_1)\right]\phi_1(x_2)\phi_2(x_1) . \qquad \text{(B.34)}$$

Here,

$$\phi_i(x) \equiv \phi^{(i)}(x) .$$

Varying the equation $a(\lambda(q); q) = 0$ by a variation of the potential q, one finds that

$$\frac{\delta\lambda}{\delta q(x)} = -\frac{1}{a_\lambda}\frac{\delta a(\lambda)}{\delta q(x)} , \qquad \text{(B.35)}$$

$$\frac{\delta^2\lambda}{\delta q(x_1)\delta q(x_2)} = -\frac{1}{a_\lambda}\left[\frac{\delta^2 a(\lambda)}{\delta q(x_1)\delta q(x_2)}\right.$$
$$\left. - \frac{d}{d\lambda}\left(\frac{1}{a_\lambda}\frac{\delta a}{\delta q(x_1)}\frac{\delta a}{\delta q(x_2)}\right)\right] , \qquad \text{(B.36)}$$

where $a_\lambda \equiv da/d\lambda$.

The expressions $\delta^2\lambda/\delta q^*(x_1)\delta q(x_2)$ and $\delta^2\lambda/\delta q^*(x_1)\delta q^*(x_2)$ are found from (B.36) by a corresponding change of q with q^*.

C. The Sine-Gordon (SG) Equation

The SG equation has the following form (*Zakharov* et al., 1980):
a) in the laboratory frame of reference:

$$\varphi_{\tau\tau} - \varphi_{\xi\xi} + \sin\varphi = 0 ; \tag{C.1}$$

b) in the variables $\xi = x + t$, $\tau = t - x$:

$$\frac{\partial^2\varphi}{\partial x\partial t} = \sin\varphi . \tag{C.2}$$

Equation (C.1) can be integrated with the use of the IST. The corresponding linear spectral problem has the form

$$\phi_x^{(1)} = i\lambda\phi^{(1)} + i\frac{\varphi_x}{2}\phi^{(2)} ,$$
$$\phi_x^{(2)} = -i\lambda\phi^{(2)} + i\frac{\varphi_x}{2}\phi^{(1)} . \tag{C.3}$$

Knowledge of the spectral data $a(\lambda)$, $b(\lambda)$ enables us to solve the Cauchy problem. The corresponding solutions will be classified as follows.

1) The coefficient $a(\lambda)$ is zero for $\lambda_1 = i\lambda$. Then, the solution is a soliton (antisoliton) of the SG equation

$$\varphi_s(x,t) = 4\tan^{-1}\exp\left[-\sigma\left(2\lambda(x - x_0) + \frac{t}{2\lambda}\right)\right] , \tag{C.4}$$

where $\sigma = \pm1$ stands for soliton (upper sign) or antisoliton (lower sign).
In the laboratory frame of reference equation (C.4) becomes

$$\varphi_s(\xi,\tau) = 4\tan^{-1}\exp\left\{\frac{\sigma[\xi - \xi_0 - v\tau]}{\sqrt{1 - v^2}}\right\}$$
$$v = \frac{4\lambda^2 - 1}{4\lambda^2 + 1} , \quad \xi_0 = (1 + v)x_0 . \tag{C.5}$$

This describes a soliton moving with velocity v. For these solutions we derive a topological charge $Q = \sigma$:

$$Q = \frac{1}{2\pi}\int_{-\infty}^{\infty} u_\xi d\xi . \tag{C.6}$$

2) Let $a(\lambda)$ have a pair of zeroes symmetric with respect to the imaginary axis

$$\lambda_1 = \mathrm{Re}\{\lambda\} + i\,\mathrm{Im}\{\lambda\} ,$$
$$\lambda_1 = -\mathrm{Re}\{\lambda\} + i\,\mathrm{Im}\{\lambda\} . \tag{C.7}$$

Then, we have a solution describing a bound state of a soliton and an antisoliton (breather):

$$\varphi_B(x,t) = 4\tan^{-1}\left\{\frac{\mathrm{Im}\{\lambda\}}{|\mathrm{Re}\{\lambda\}|}\frac{\sin\left[\frac{\mathrm{Re}\{\lambda\}}{2|\lambda|^2}t - 2\mathrm{Re}\{\lambda x\} - \varphi_0\right]}{\cosh\left[2\mathrm{Im}\{\lambda(x - x_0)\} + \frac{\mathrm{Im}\{\lambda\}}{2|\lambda|^2}t\right]}\right\}. \tag{C.8}$$

In the laboratory frame of reference

$$\varphi_B(\xi,\tau) = 4\tan^{-1}\left\{\frac{\mathrm{Im}\{\lambda\}}{\mathrm{Re}\{\lambda\}}\frac{\sin\left[\frac{\mathrm{Re}\{\lambda\}}{|\lambda|}\left(\frac{\tau - v\xi}{\sqrt{1-v^2}} - \phi_0\right)\right]}{\cosh\left[\frac{\mathrm{Im}\{\lambda\}}{|\lambda|}\left(\frac{\xi - v\tau - \xi_0}{\sqrt{1-v^2}}\right)\right]}\right\}.$$

The topological charge Q for (C.7) is zero.

The energy is

$$E = \frac{1}{2}\int_{-\infty}^{\infty}dx\left(\varphi_t^2 + \varphi_x^2 + 4\sin^2\frac{\varphi}{2}\right). \tag{C.9}$$

For a soliton (or, as it is usually called, the kink) this is

$$E_s = 8\sqrt{1 - v^2}, \tag{C.10}$$

and for the breather, we find

$$E_b = 16(1 - v^2)^{-1/2}\sin\gamma, \quad \gamma = \tan^{-1}\left[\frac{\mathrm{Im}\{\lambda\}}{\mathrm{Re}\{\lambda\}}\right].$$

Canonically, conjugate variables have the form

$$P_\lambda = \frac{4}{\pi\lambda}\ln\frac{1}{|a(\lambda)|^2}, \quad Q_\lambda = \arg b(\lambda).$$

In terms of the canonical variables, the Hamiltonian for the SG equation is

$$H = \int_0^\infty\left(\lambda + \frac{1}{4\lambda}\right)P_\lambda d\lambda + \sum_k\frac{2\left(e^{-P_k} + 4e^{P_k}\right)}{\gamma}$$

$$+ \sum_k\frac{4}{\gamma}\sin\frac{\eta_k\gamma}{16}\left(e^{-n_k/4} + 4e^{n_k/4}\right), \tag{C.11}$$

$$\eta_k = \frac{16}{\gamma}\arg\lambda_k; \quad n_k = 4\ln|\lambda_k|.$$

The first term in this expression is the contribution from the continuous spectrum, the second is that from the solitons, and the third term is the contribution from the breathers. Such a form for the total energy is useful when calculating the energy of waves radiated by solitons and breathers.

The Hamiltonian of a breather in terms of the variables γ, θ (where θ_τ is the breather self-frequency) is

$$H_b = \sin\gamma. \tag{C.12}$$

For $\gamma > \gamma_{sep} = \pi/2$ the breather is separated into a free kink and antikink pair.

We next consider the equations of perturbation theory for solitons (*McLaughlin* and *Scott*, 1983). A perturbed SG equation is

$$\varphi_{tt} - \varphi_{xx} + \sin \varphi = \varepsilon R(\varphi) .$$

In the adiabatic approximation, the one-soliton solution is taken in the form

$$\varphi_s(x, t) = 4 \tan^{-1} \left\{ \exp \frac{x - \zeta(t)}{\sqrt{1 - v^2}} \right\} . \tag{C.13}$$

To first order in ε, the velocity v and the coordinate of the soliton centre $\zeta(t)$ satisfy

$$\frac{dv}{dt} = -\frac{\varepsilon \sigma}{4} (1 - v^2)^{1/2} \int_{-\infty}^{\infty} dz \, R(\varphi_s(z)) \operatorname{sech} z , \tag{C.14}$$

$$\frac{d\zeta}{dt} = v - \frac{\varepsilon \sigma}{4} (1 - v^2) \int_{-\infty}^{\infty} dz \, z R(\varphi_s(z)) \operatorname{sech} z , \tag{C.15}$$

$$z = \frac{x - \zeta}{\sqrt{1 - v^2}} , \qquad \sigma = \pm 1 .$$

The equations for the breather parameters are (*Karpman* et al., 1983; *Kivshar* and *Kosevich*, 1982):

$$\frac{d\gamma}{dt} = \varepsilon (1 - v^2)^{1/2} (4 \cos \gamma)^{-1} I_1 , \tag{C.16}$$

$$\frac{dv}{dt} = -\varepsilon (1 - v^2)^{3/2} (4 \cos \gamma)^{-1} I_2 , \tag{C.17}$$

$$\frac{d\theta}{dt} = \cos \gamma (1 - v^2)^{1/2} - \varepsilon (1 - v^2)^{1/2} [v \cot \gamma I_3$$
$$+ \cos^2 \gamma (1 - v^2) I_4] - I_5 (4 \sin \gamma \cos^2 \gamma)^{-1} , \tag{C.18}$$

$$\frac{dx_0}{dt} = v + \varepsilon (1 - v^2)(I_3 - v \tan \gamma I_4)(2 \sin \gamma)^{-2} , \tag{C.19}$$

$$I_1 = \int_{-\infty}^{\infty} \frac{\cosh z \sin \Phi}{\cosh^2 z + A^2} R(\varphi_B(z)) dz ,$$

$$I_2 = \int_{-\infty}^{\infty} \frac{\sinh z \cos \Phi}{\cosh^2 z + A^2} R(\varphi_B(z)) dz ,$$

$$I_3 = \int_{-\infty}^{\infty} \frac{z \cosh z \sin \Phi}{\cosh^2 z + A^2} R(\varphi_B(z)) dz ,$$

$$I_4 = \int_{-\infty}^{\infty} \frac{\cosh z \cos \Phi}{\cosh^2 z + A^2} R(\varphi_B(z)) dz ,$$

$$I_5 = \int_{-\infty}^{\infty} \frac{z^2 \cosh z \cos \Phi}{\cosh^2 z + A^2} R(\varphi_B(z)) dz ,$$

$$A = \tan \gamma \cos \varphi \, ,$$

$$\Phi = \frac{\text{Re } \lambda}{|\lambda|} \frac{t - vx}{\sqrt{1 - v^2}} + \Phi_0 \, .$$

The Jost functions for the one-soliton solution are

$$\varphi = \frac{e^{-ikz}}{\lambda + i\nu} \begin{pmatrix} -\sigma\nu \text{ sech } z \\ \lambda - i \tanh z \end{pmatrix} \, ,$$

$$\phi = \frac{e^{ikz}}{\lambda + i\nu} \begin{pmatrix} \lambda + i\nu \text{ sech } z \\ \sigma\nu \text{ sech } z \end{pmatrix} \, . \qquad (C.20)$$

Here, $k = 1/2[\lambda - (1/4\lambda)]$.

D. Finite-Gap Solutions in NLS Equation

Here, we briefly describe explicit formulae for the finite-gap solutions of the NLS equation.

1) Finite-gap potentials: We consider the spectral problem for the operator

$$L = \begin{pmatrix} -i & 0 \\ 0 & i \end{pmatrix} \frac{d}{dx} + \begin{pmatrix} 0 & iu(x) \\ -i\bar{u}(x) & 0 \end{pmatrix}$$

where u is a periodic function with period T. The spectrum of this operator has a zone structure, and boundaries of these zones are eigenvalues for either the periodic or antiperiodic problem for the equation $Ly = \lambda y$. The potential u is called a finite-gap potential if the spectrum of L has only a finite number of gaps. It turns out that the evolution of potentials, according to the NLS equation, preserves the property that potentials remain finite-gap.

The set of all finite-gap potentials is a sum of separate finite-dimensional families. All these families can be described explicitly by $1/m$; there are formulae describing all finite-gap potentials. Moreover, these families are invariant with respect to the NLS equation, so one has a corresponding set of finite dimensional dynamical systems. All these systems can be solved: the explicit formulae for finite-gap solutions of the NLS equation can be presented.

Let a potential u have n gaps, then one can associate with it a Riemann surface Γ of genus n, and n points μ_j of this surface. There exists a degree-two covering $\Gamma \to \mathbf{P}^1$ of the Riemann sphere. The covering set is the points $\lambda_1, \bar{\lambda}_1, \ldots, \lambda_{n+1}, \bar{\lambda}_{n+1}$, i.e., Γ is a surface for the function $\sqrt{\Pi(\lambda - \lambda_1)(\lambda - \bar{\lambda}_1)}$. The points $\lambda_i, \bar{\lambda}_i$ are the boundaries of zones of the spectrum. Also, μ_j are the poles of the Floquet solution of L. It appears that μ_j depends neither on x nor t. The Floquet solution is uniquely determined by its poles, zeroes, and asymptotics. Only the zeroes $\mu_j(x, t)$ vary, and one can find equations that describe the variation of $\mu_j(x, t)$. These equations can be solved using the Abel-Jacobi mapping. Thus, one obtains "explicit" formulae for the solutions in terms of the Riemann theta function, using standard techniques from the theory of Riemann surfaces.

2) Explicit formulae: Let u be a solution with n gaps, and $\lambda_1, \bar{\lambda}_1, \ldots, \lambda_{n+1}, \bar{\lambda}_{n+1}$ be the boundaries of the zones. All but two constants in the formulae depend only on λ_i. The other two depend on λ_i and some additional data.

Unfortunately, one must start with many definitions. The formula for the finite gap NLS equation is

$$u(x, t) = \frac{\theta(g - r)}{\theta(g + r)} \exp(iE_0 x + iN_0 t + A_0) ,$$

where θ is a Riemann theta function. It is constructed by using the symmetric matrix $B = (b_{jj})$ (with positively defined imaginary part) and depends on $z_1, \ldots, z_n$:

$$\theta(z) = \sum \exp\{\pi i[(Bm, m) + 2(z, m)]\} \,,$$

where m spans the set of vectors whose coordinates are integer; $g = g(x, t) = \gamma x + \alpha t + \ell$; γ, α, ℓ are described below, and r, E_0, N_0, A_0 are constants.

Now, we define B, γ, α, ℓ, r, E_0, N_0, A_0. To do this we need standard cycles of integration a_i, b_i on Γ. Γ is a surface of $\sqrt{\Pi(\lambda - \lambda_i)(\lambda - \bar\lambda_i)}$ and is constructed by cutting two copies of $\mathbb{C}$ along $[\lambda_i, \bar\lambda_i]$ and "glueing" these together. Cycle a_i goes around $[\lambda_i, \bar\lambda_i]$ on the upper copy of $\mathbb{C}$; b_i goes from λ_i to λ_n on the upper copy of $\mathbb{C}$ and returns to λ_i on the lower copy (see Sect. 3.4 for an example). We define the matrix (a_{ij}):

$$a_{ij} = \int_{a_i} \frac{\lambda^i d\lambda}{\sqrt{P(\lambda)}} \,,$$

$$(P(\lambda) = \Pi(\lambda - \lambda_i)(\lambda - \bar\lambda_i)) \,,$$

the matrix (c_{ij}):

$$(c_{ij}) = (a_{ij})^{-1} \,,$$

and the differentials U_i:

$$U_i = \left(\sum_j c_{ij}\lambda^j\right) \frac{d\lambda}{\sqrt{P(\lambda)}} \,.$$

We have

$$\int_{a_i} U_j = \delta_{ij} \,,$$

and we set

$$b_{ij} = \int_{b_i} U_j \,,$$

$$\gamma_i = 2ic_{i,n-2} \,,$$

$$\alpha_i = -4ic_{i,n-2}\frac{\sum(\lambda_i - \bar\lambda_i)}{2} + c_{i,n-3} \,,$$

$$\ell_i = \sum_k U_i[\mu_k(0,0)] + \frac{1}{2}\sum_j b_{ij} - \frac{i}{2} \,,$$

where $U_i(\lambda) = \int_{\lambda_n}^{\lambda} U_i$. One can choose arbitrary points $\mu_k(0,0)$, subject to one condition which must be satisfied: these points must lie over prohibited zones.

Further,

$$r_i = U_i(\infty^+) = \int_{\lambda_n}^{\infty^+} U_i \,,$$

where ∞^+ is an ∞-point on the upper leaf of Γ,

$$E_0 = \sum_i (\lambda_I + \bar{\lambda}_I) - \sum_i \int_{a_i} \lambda U_i \ ,$$

$$N_0 = 4 \sum_i \int_{a_i} \lambda^2 U_i - \sum_i (\lambda_i + \bar{\lambda}_i) - 4R_1 \ .$$

R_1 is defined by the following condition: let ω be a differential with $\int_{a_i} \omega = 0$, then and only then the poles are of the second order at $\infty^\pm$; then

$$\omega|_{\lambda \to \infty^\pm} = \pm(\lambda^2 + R_1 + \ldots) \ .$$

3) Degenerate case: When $g = 3$, the spectral pattern consists of eight points a, $\bar{a}$, c, $\bar{c}$, λ_0, $\bar{\lambda}_0$, λ_1, $\bar{\lambda}_1$. We want to see what will happen to the formulae if $\lambda_1 \to \lambda_0$.

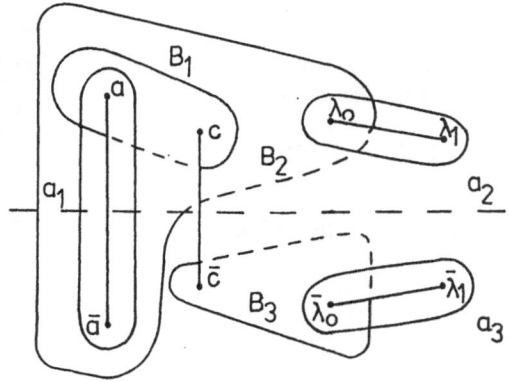

Fig. D.1. Cycles of integration in the complex λ-plane for the degenerate case $g = 3$

We choose a-cycles and b-cycles as shown in Fig, D.1, where the dashed parts of the curves belong to the lower sheet of the surface. Let these cycles be degenerate $(\lambda_1 \to \lambda_0)$ (Fig. D.2). The differentials U_i are uniquely determined by condition $\int_{a_i} U_j = \delta_{ij}$ and become

$$U_1 = \frac{d\lambda}{\sqrt{P(\lambda)}} \left[2 \int_{\bar{a}}^a \frac{d\lambda}{\sqrt{P(\lambda)}} \right]^{-1} ; \quad P(\lambda) = (\lambda - a)(\lambda - \bar{a})(\lambda - c)(\lambda - \bar{c}) \ ,$$

$$U_2 = \frac{1}{2\pi i} \frac{d\lambda}{(\lambda - \lambda_0)\sqrt{P(\lambda)}} \sqrt{P(\lambda_0)} - U_1 \frac{\sqrt{P(\lambda_0)}}{\pi i} \int_{\bar{a}}^a \frac{d\lambda}{(\lambda - \lambda_0)\sqrt{P(\lambda)}} \ ,$$

$$U_3 = \frac{1}{2\pi i} \frac{d\lambda}{(\lambda - \bar{\lambda}_0)\sqrt{P(\lambda)}} \sqrt{P(\bar{\lambda}_0)} - U_1 \frac{\sqrt{P(\bar{\lambda}_0)}}{\pi i} \int_{\bar{a}}^a \frac{d\lambda}{(\lambda - \bar{\lambda}_0)\sqrt{P(\lambda)}} \ ,$$

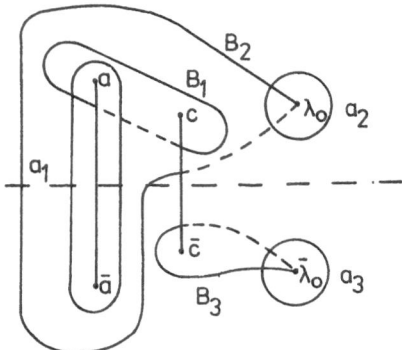

Fig. D.2. Degeneracy of integration cycles

Note that U_i are holomorphic outside $\lambda_0, \bar{\lambda}_0$, where they can only have poles of order 1; the condition $\int_{a_i} U_1 = 0$, $i = 2, 3$ means that U_i is holomorphic, and $\int_{a_i} U_j = \delta_{ij}$ means that the residue of U_j at λ_0 ($i = 2$) is equal to $1/2\pi i$.

Thus, U_1 is regular, U_2 has a pole at λ_0, $\bar{U}_2$ has a pole at $\bar{\lambda}_0$, and so, we deduce that $b_{22} = \int_{b_2} U_2$ and $b_{33} = \int_{b_3} U_3$ are infinite, or to be more precise, that their imaginary part tends to $+\infty$. This means that in the sum

$$\theta(g(x,t) \pm r) = \sum_m \exp\left\{\pi i[(Bm, m) + 2(g \pm r, m)]\right\}$$

the non-zero terms appear only at the following values of m:

1) $m_2 = m_3 = 0$,
2) $m_2 = -1$, $\quad m_3 = 0$;
3) $m_2 = 0$, $\quad m_3 = -1$;
4) $m_2 = m_3 = -1$.

Recall that $g(x,t) = \gamma x + \alpha t + \ell$, and ℓ_i contains $1/2 \sum_j b_{ji}$, so that only the listed terms do not contain b_{22} and b_{33}. Thus, we have

$$\check{\theta}(p) := \lim \theta(p) = \sum_{m_1} \exp\{\pi i[b_{11}m_1^2 + 2p_1 m_1]\}+$$

$$+ \sum_{m_1} \exp\{\pi i[b_{11}m_1^2 - 2b_{21}m_1 + 2p_1 m_1 - 2p_2]\}+$$

$$+ \sum_{m_1} \exp\{\pi i[b_{11}m_1^2 - 2b_{31}m_1 + 2p_1 m_1 - 2p_3]\}+$$

$$+ \sum \exp\{\pi i[b_{11}m_1^2 - 2(b_{21} + b_{31})m_1 + 2b_{32} + 2p_1 m_1 - 2(p_2 + p_3)]\} .$$

The first sum is (by definition) a Jacobi theta function with the period equal to

$$b_{11} = 2 \int_{\bar{c}}^a U_1 .$$

Finally, we can replace b_{31} by $\bar{b}_{21}$; indeed,

$$b_{21} = 2 \int_{\bar{c}}^{\lambda_0} U_1 \,,$$

$$b_{31} = 2 \int_{\bar{c}}^{\bar{\lambda}_0} U_1 \,,$$

but $2 \int_{\bar{c}}^{\lambda_0} U_1 = 2 \int_c^{\lambda_0} U_1 + 2 \int_{\bar{c}}^c U_1 = \bar{b}_{31} - 2 \int_{\bar{a}}^a U_1 = \bar{b}_{31} - 1$ (recall that $2 \int_{\bar{a}}^a U_1 = 1$ and $\theta(p+1) = \theta(p)$.

E. The Tensor Product of Two Matrixes

The tensor product of two $n \times n$ matrixes A and B is defined by

$$M^{ij}_{kp} = (A \otimes B)^{ij}_{kp} = A_{ij} B_{kp}, \quad i,j,k,p = 1,2,3 \ldots n. \tag{E.1}$$

Here, M is the matrix of the $n^2 \times n^2$ dimension. The matrix elements are labelled by the block indices i, j, and internal indices k, p. For example, in the case $n = 2$:

$$M = (A \otimes B) = \begin{pmatrix} A_{11} B & A_{12} B \\ A_{21} B & A_{22} B \end{pmatrix}$$

$$= \begin{pmatrix} A_{11} B_{11} & A_{11} B_{12} & A_{12} B_{11} & A_{12} B_{12} \\ A_{11} B_{21} & A_{11} B_{22} & A_{12} B_{21} & A_{12} B_{22} \\ A_{21} B_{11} & A_{21} B_{12} & A_{22} B_{11} & A_{22} B_{12} \\ A_{21} B_{21} & A_{21} B_{22} & A_{22} B_{21} & A_{22} B_{22} \end{pmatrix}. \tag{E.2}$$

The matrix product of two $n^2 \times n^2$ matrices M and N is the $n^2 \times n^2$ dimension of matrix L

$$L^{ij}_{kp} = (M\, N)^{ij}_{kp} = \sum_{m,\ell} M^{im}_{k\ell} N^{mj}_{\ell p}. \tag{E.3}$$

The unit matrix of the dimension $n^2 \times n^2$ has the form

$$\hat{E} = I \otimes I, \quad \hat{E}^{ij}_{kp} = \delta_{ij} \delta_{kp}. \tag{E.4}$$

The matrix permutation of $n^2 \times n^2$ is defined by an equality

$$\hat{p}(f \otimes \varphi) = \varphi \otimes f, \tag{E.5}$$

where f and φ are the vectors in the space of dimension n.

$$\hat{P}^{ij}_{kp} = \delta_{ip} \delta_{kj}, \quad \hat{p}^2 = \hat{E}. \tag{E.6}$$

For any matrices A and B with commutative elements the following equality is true:

$$\hat{P}(A \otimes B)\hat{P} = B \otimes A. \tag{E.7}$$

In the case when $n = 2$, the matrix $\hat{P}$ is expressed through the Pauli matrix:

$$\hat{P} = \frac{1}{2} \left(\hat{E} + \sum_{a=1}^{3} \sigma_a \otimes \sigma_a \right), \tag{E.8}$$

and has the form

$$\hat{P} = \begin{pmatrix} 1 & 0 & 0 & 0 \\ 0 & 0 & 1 & 0 \\ 0 & 1 & 0 & 0 \\ 0 & 0 & 0 & 1 \end{pmatrix}. \tag{E.9}$$

The Poisson bracket (PB) of a tensor product of two matrices A and B is defined as

$$\{A \overset{\otimes}{,} B\} = i \int \left[\frac{\delta A}{\delta \phi(x)} \otimes \frac{\delta B}{\delta \phi^*(x)} - \frac{\delta A}{\delta \phi^*(x)} \otimes \frac{\delta B}{\delta \phi(x)} \right] dx \ ,$$

or

$$\{A \overset{\otimes}{,} B\}^{ij}_{k\ell} = \{A_{ij}, B_{k\ell}\} \ ,$$

i.e., the Poisson bracket is defined between the elements of A and B matrices.

In tensor notation the general properties of the Poisson brackets, i.e., the properties of antisymmetric (6.5) and Jacobi identity (6.6) takes the respective form:

$$\{A \overset{\otimes}{,} B\} = -\hat{P}\{B \otimes A\}\hat{P} \ , \tag{E.12}$$

$$\{A \overset{\otimes}{,} \{B \overset{\otimes}{,} C\}\} + \hat{P}_{13}\hat{P}_{23}\{C \overset{\otimes}{,} \{A \overset{\otimes}{,} B\}\}\hat{P}_{23}\hat{P}_{13}$$
$$+ \hat{P}_{13}\hat{P}_{12}\{B \overset{\otimes}{,} \{C \overset{\otimes}{,} A\}\}\hat{P}_{12}\hat{P}_{13} = 0 \ , \tag{E.13}$$

where the matrix indices $\hat{P}$ denote numbers of factors of the tensor product of three matrices on which an operator $\hat{P}$ is acting. For example, $\hat{P}_{12}$ denotes a matrix coinciding with the matrix P in a product of the first two factors.

References

(Numbers in square brackets like [4.5] refer to the section in which these references are cited.)

Abdullaev F. Kh. (1982), Izv. VUZov–Radiofizika 25, 756 (in Russian) [4.5]

Abdullaev F. Kh. (1983), Pisma Zh. Tekh. Fiz. 9, 305 (in Russian) [4.5]

Abdullaev F. Kh. (1983), Lebedev Institute Reports (Moscow) 1, 51 (in Russian) [4.7]

Abdullaev F. Kh., Abdullaev S. S. (1980), Izv. VUZov–Radiofizika 23, 789 (in Russian) [7]

Abdullaev F. Kh., Darmanyan S. A., Umarov B. A. (1985), Phys. Lett., A108, 51 [4.7]

Abdullaev F. Kh., Umarov B. A. (1985), Izv. VUZ'ov–Radiofizika, 28, 664 (in Russian) [4.7]

Abdullaev F. Kh., Khabibullaev P. Kh. (1986), *Dynamics of Solitons in Inhomogeneous Condensed Matter*, FAN, Tashkent (in Russian) 2.6 [2.6, 4.5]

Abdullaev F. Kh., Darmanyan S. A., Abrarov R. M. (1986), "Evolution of Ultrashort Intense Pulses in Optical Waveguides", in: *Technology and Properties of Optical Fibers*, FAN, Tashkent, 127–136 [2.6]

Abdullaev F. Kh., Darmanyan S. A., Djumaev M. R. (1986), Izv. Akad. Nauk UzSSR 6, 21 (in Russian) [4.6]

Abdullaev F. Kh., Darmanyan S. A., Khabibullaev P. K. (1987), *Optical Solitons*, FAN, Tashkent (in Russian) [1]

Abdullaev F. Kh., Darmanyan S. A. (1988), "Evolution of Random Field in Integrable and Almost-Integrable Systems", in: *Nonlinear and Turbulent Processes in Physics*, Naukova Dumka, Kiev, 3–7 [4.4]

Abdullaev F. Kh., Tadjimuratov S., Tartakovsky G. (1989), Izv. Akad. Nauk UzSSR 1, 15 (in Russian) [5.6]

Abdullaev F. Kh., Abrarov R. M., Darmanyan S. A. (1989), Opt. Lett. 14, 131 [3.5]

Abdullaev F. Kh. (1989), Phys. Rep. 179, 1 [4.7]

Abdullaev F. Kh., Darmanyan S. A. (1989), Radiofizika 32, 1048 (in Russian) [4.3, 4]

Abdullaev F. Kh., Darmanyan S. A., Djumaev M. R. (1989), Phys. Lett. A141, 423 [4.6]

Abdullaev F. Kh., Tadjimuratov S. (1991), Dokl. Akad. Nauk SSSR (in Russian) [5.2]

Abdullaev F. Kh., Abrarov R. M., Darmanyan S. A. (1991), "Solitons Dynamics in Directional Couplers", in: *Digest of Nonlinear Guided-Wave Phenomena*. Ed. Boadman M., Cambridge, 3 [3.5]

Abdullaev F. Kh., Darmanyan S. A., Goncharov V. (1992), Pisma JTP 18, 21 (in Russian) [3.5]

Abdullaev F. Kh., Gulyamov R. (1992), Pisma JTP 18, 10 (in Russian) [3.5]

Abdumalikov A. A., Abdullaev F. Kh., Stamenkovich L. (1983), Phys. Stat. Solid(b) 120, K33 [4.5]

Ablowitz M. J., Kaup D. J., Newell A. C. (1976), J. Math. Phys. 15, 1852 [5.1]

Ablowitz M. J., Segur H. (1981), *Solitons and the Inverse Scattering Transform*, SIAM, Philadelphia [2.3, 4.1]

Abrarov R. M., Darmanyan S. A. (1985), Izv. Akad. Nauk UzSSR 6, 45 (in Russian) [3.3, 4.5, 5.5]

Abrarov R. M. (1990), Ph. D. thesis, Thermophysics Department of Uzbek Acad. Sci., Tashkent (in Russian) [3.5]

Acevec A.B., Wabnitz (1992), Opt. Lett. 17, 25 [2.13]

Adams M. (1981), *An Introduction to Optical Waveguides*, Wiley, New York [App. A]

Agrawal G. P., Potasek M. J. (1986), Phys. Rev. Lett. 33, 1765 [2.1]

Akhmanov S. A., Sukhorukov A. P., Khokhlov R. V. (1967), Usp. Fiz. Nauk 93, 19 (English transl. Sov. Phys. Usp. 93, 609, 1968) [1]

Akhmanov S. A., Vysloukh V. A., Chirkin A. S. (1986), Usp. Fiz. Nauk **149**, 449 (in Russian) [1, 4.1]

Akhmediev N. N. (1989), DSc (phys.math.) thesis, Moscow State University (in Russian) [2.9]

Akhmediev N. N., Eleonsky V. M., Kulagin N. N. (1985), Zh. Eksp. Theor. Fiz. **89**, 1542 (English transl. Sov. Phys. JETP, 1985, 62, 894) [2.9]

Allen L., Eberley J. (1975), *Optical Resonance and Two-Level Atoms*, Wiley, New York [5.1], [6.4]

Anderson D. (1983), Opt. Commun. **48**, 107 [2.1]

Anderson D., Lisak M. (1983), Phys. Rev. **A27**, 1393 [2.5]

Anderson D., Lisak M. (1984), Opt. Lett. **9**, 468 [2.8]

Anderson D., Lisak M. (1986), Opt. Lett. **11**, 174 [2.8, 3.1]

Andreev A. V., Emel'yanov V. I., Il'inskii Yu. A. (1978), Sov. Phys. Usp. **23**(8), 493 [6.4]

Andrushko L. M., Karpluk K. S., Ostrovsky S. B. (1987), Radiotekh. Elektron. **32**, 427 (in Russian) [3.5]

Aranson I. S., Gorshkov K. A., Rabinovich M. I. (1984), Zh. Eksp. Teor. Fiz. **84**, 929 (in Russian) [4.7]

Arnold V. I. (1983), *Introduction to Catastrophe Theory*, Nauka, Moscow (in Russian) [5.8]

Aueung J., Yariv A. (1978), IEEE J. Quant. Electr. **14**, 347 [2]

Azimov B. S., Isaev S. K., Luzgin S. N., Trukhov D. V. (1986), Izv. Akad. Nauk SSSR, ser. fiz. **50**, 2268 (in Russian) [4.1]

Basharov A. M., Maimistov A. I. (1986), *Photonics. Nonlinear Coherent Processes*. Preprint Moscow Phys. Eng. Inst. No.18, Moscow (in Russian) [5.1]

Bass F. G., Kivshar Yu. S., Konotop V. V. (1986a), Preprint Radio-Electronics Inst., Akad. Nauk SSSR, No. 296, 36p, Kharkov (in Russian) [4.3]

Bass F. G., Kivshar Yu. S., Konotop V. V. (1986b), Z. Phys. **B65**, 209 [4.5]

Bass F. G., Kivshar Yu. S., Konotop V. V., Puzenko S. A. (1989), Opt. Comm. **68** , 385 [4.3]

Bass F. G., Kivshar Yu. S., Konotop V. V., Pritula G. M. (1989), Opt. Comm. **70**, 309 [4.5, 7]

Belanger P. A. (1988), J. Opt. Soc. Am. **B5**, 798 [2.10]

Belanger P. A. (1991), JOSA **B8**, 2077 [2.11]

Bendow B. (1980), J. Opt. Soc. Am. **70**, 539 [2.1, 9]

Benjamin T. B., Feir J. E. (1967), J. Fluid Mech. **27**, 417 [2.8]

Berg P., If F., Christiansen P., Skovgaard D. (1987), Phys. Rev. **A35**, 4167 [2.11]

Bespalov V. I., Talanov V. I. (1966), Pisma Zh. Eksp. Teor. Fiz. **3**, 471 (in Russian) [2.9]

Bikbaev R. F., Bobenko A., Its A. R. (1984), Preprint Phys. Techn. Inst., Donetzk, No. 84–6 (in Russian) [3.4]

Bishop A. R., Fesser K., Lomdahl P. S., Kerr W. C., Williams M. B., Trullinger S. E. (1983), Phys. Rev. Lett. **50**, 1095 [4.7]

Blow K. J., Doran N. J. (1983), Electr. Lett. **19**, 429 [3.2]

Blow K. J., Doran N. J., Cummins E. (1983), Opt. Commun. **48**, 107 [2.1]

Blow K. J., Doran N. J. (1984), Phys. Rev. Lett. **52**, 526 [4.8]

Blow K. J., Wood D. (1986), IEEE J. Quant. Electron. **22**, 1109 [1]

Blow K. J., Doran N. J., Wood D. (1987), Opt. Lett. **12**, 202 [2.13]

Bogolyubov N. M., Korepin V. E. (1985), Teor. Mat. Fiz. **64**, 92 (in Russian) [6.1]

Bolshov L. A., Likhansky V. V. (1985), Kvant. Elektron. **12**, 1339, (in Russian) [1, 5.3, 6.2]

Bullough R. K., Jack P. M., Kitchenside P. W., Saunders R. (1979), Phys. Scripta **20**, 364 [5.1, 6.3]

Bullough R. K., Saunders R., Feuillade C. (1978), in: *Coherence and Quantum Optics IV*, ed. Mandel L., Wolf E., Plenum, London, 26 [6.3]

Burlak G. N., Gryshalsky V. V., Kotzarenko N. Ya. (1986), Radiofizika **29**, 1259 (English transl.: Radiophys. and Quant, Electr. **29**, 959, 1986) [7]

Burzlaff J. (1988), J. Phys. A, Math. Gen. **21**, 561 [2.3]

Caglioti E., Trillo S., Wabnitz S., Crosigniani B., Di Porto P. (1989), "Soliton Dynamics in Bimodal Optical Fibers", in: *Nonlinear Wave Phenomena in Waveguides*, Conf. digest. Eds. Stegeman G., Stolen R. (Houston, Texas), 124–127 [3.3]

Campbell D. K., Peyrard M., Sodano P. (1986), Physica **D19**, 165 [3.1]

Carter S. J., Drummond P. D. (1991), Phys. Rev. Lett. **67**, 3757 [6]

Chirikov B. V. (1979), Phys. Rep. **52**, 278 [4.7]

Chirikov B. V., Shepelyansky Yu. (1982), JETP Lett. **34**, 590 [2.12]

Christiansen P. L. (1986), Opt. Commun. **57**, 350 [2.11]

Christodoulides D. N., Joseph R. I. (1984), Opt. Lett. **9**, 408 [2.3]

Christodoulides D. N., Joseph R. I. (1988), Opt. Lett. **13**, 53 [2.13]

Christodoulides D. N., Joseph R. T. (1989), Phys. Rev. Lett. **62**, 1746 [2.13]

Chu F., Scott A. C. (1975), Phys. Rev. A**12**, 2060 [5.2]

Chu P. L., Desem C. (1984), Electr. Lett. **53**, 218 [3.3]

Cotter D. (1983), J. Opt. Commun. **4**, 10 [2]

Crossiniani B., Cutolo A., DiPorto P. (1982), J. Opt. Soc. Am. **72**, 1136 [2.10]

Darmanyan S. A. (1992), Opt. Commun. **90**, 301 [4.6]

de Angelis C., Matua F., Wabnitz S. (1992), Opt. Lett. **17**, 650 [4.5]

Demchuk M. I., Mikhailov V. P., Prokhorov A. M., Sisakyan I. N., Shvartzburg A. B. (1984), Pisma Zh. Tekhn. Fiz. **15**, 338 (in Russian) [2]

Desaix M., Anderson D., Lisak M. (1990), Opt. Lett. **15**, 18 [2.6]

Desem C., Chu P.L. (1987a), Opt. Lett **12**, 349 [2.2]

Desem C., Chu P.L. (1987b), Electron. Lett. **23**, 260 [3.4]

Dianov E. M., Isaev S. K., Kornienko L. S., Kravtzov N. V., Firsov N. V. (1979), Izv. Akad. Nauk SSSR **43**, 266 [2]

Dianov E. M., Zakhidov E. A., Karasik A. Ya. (1983), Pisma Zh. Tekhn. Fiz. **9**, 1455 (in Russian) [2]

Dianov E. M., Prokhorov A. M., Serkin V. N. (1983), Dokl. Akad. Nauk SSSR **273**, 1112 (English transl.: Sov. Phys. -Dokl. **28**, 1036, 1983) [2, 1]

Dianov E. M., Karasik A. Ya., Prokhorov A. M., Serkin V. N. (1984), Izv. Akad. Nauk SSSR **48**, 1458 (in Russian) [2.1]

Dianov E. M., Nikonova Z. S., Prokhorov A. M., Serkin V. N. (1985), Dokl. Akad. Nauk SSSR **283**, 1342 (in Russian) [2.7, 4.6]

Dianov E. M., Prokhorov A. M. (1986), Usp. Fiz. Nauk **148**, 289 (in Russian) [1]

Dianov E. M., Grudinin A. B., Khaidarov D. V., Korobkin D. V., Prokhorov A. M., Serkin V. N. (1989), Fib. Integ. Optics **8**, 61 [2.7]

Dianov E. M., Mamyshev P. V., Prokhorov A. M., Chernikov S. V. (1990), Opt. Lett. **14**, 18 [2.2]

Dianov E. M., Luchnikov A. V., Pilipetskii A. N., Starodumov A. N. (1990), Opt. Lett. **15**, 314 [3.2]

Doran N. J., Wood D. (1988), Opt. Lett. **13**, 56 [2.13]

Druhl K., Wentzel R. G., Carsten J. L. (1983), Phys. Rev. Lett. **51**, 1171 [5.6]

Druhl K., Carsten J. L., Wentzel R. G. (1985), J. Stat. Phys. **39**, 615 [5.6]

Elgin J.N. (1985) Phys. Lett. **110A**, 441 [4.4]

Elyutin S., Maimistov A., Manykin E. (1989), Phys. Lett. A**132**, 25 [5.1]

Emplit P., Hamaide J. P., Reynand F., Froely C., Barthelemy A. (1987), Opt. Comm. **62**, 374 [2.4]

Englund J., Bowden C. M. (1986), Phys. Rev. Lett. **57**, 2661 [5.6]

Faddeev L. D. (1979), "Quantum Integrable Systems" in: *Problems of the Quantum Field Theory*, Dubna, 249 (in Russian) [6.1]

Faddeev L. D. (1980), Sov. Sci. Rev. C**1**, 107 [6.2]

Faddeev L. D. (1980), in: *Solitons*, eds. Bullough R. K., Caudrey P. S., Springer, Heidelberg [6.1]

Faddeev L. D., Takhtadjan L. A. (1986), *Hamiltonian Approach in the Soliton Theory*, Nauka, Moscow, 527p. (in Russian) [6.1]

Fattakhov A., Chirkin A. S. (1983), Kvant. Elektron. **10**, 1989 (in Russian) [4.2]

Fattakhov A., Chirkin A. S. (1984), Kvant. Elektron. **12**, 2349 (in Russian) [4.2]

Friberg S. R. (1992), Opt. Lett. **16**, 1484 [3.5]

Friberg S. R., DeLong K. W. (1992), Opt. Lett. **17**, 979 [3.5]

Gabitov I. P., Zakharov V. E., Mikhailov A. V. (1984), Zh. Eksp. Teor. Fiz. **86**, 1204 (English transl.: Sov. Phys. JETP **59**, 703, 1984) [5.4]

Gabitov I. R., Zakharov V. E., Mikhailov A. V. (1985), Teor. Mat. Fiz. **63**, 11 (in Russian) [5.4]

Gabitov I. R., Romagnoli M., Wabnitz S. (1991), Appl. Phys. Lett. **16**, 1811 [5.1]

Gambriquas J. M. (1983), Opt. Lett. **8**, 183 [2]

Gardner C. S., Green J. M., Kruskal M. D., Miura R. M. (1967), Phys. Rev. Lett. **19**, 1095 [1]

Golovchenko E. A. et al. (1986), Dokl. Akad. Nauk SSSR **288**, 851 (in Russian) [2.6]

Gordon J. P. (1983), Opt. Lett. **8**, 596 [3.2]

Gordon J. P. (1986), Opt. Lett. **11**, 662 [2.12, 3.2]

Gordon J. P., Haus H. A. (1986), Opt. Lett. **11**, 665 [4.6]

Gorshkov K. G., Ostrovsky L. A. (1981), Physica D3, 341 [3.1]

Gouveia-Neto A. S., Gomes A. S., Taylor J. R. (1987), Opt. Lett. 12 **11**, 927 [2.7]

Gouveia-Neto A. S., Gomes A. S., Taylor J. R. (1988), IEEE J. Quant. Electr. **24**, 332 [2.7]

Gredeskul S. A., Kivshar Yu. S. (1989), Phys. Rev. Lett. **62**, 977 [2.3]

Gredeskul S. A., Kivshar Yu. S., Yanovskaya M. V. (1990), Phys. Rev. A41, N7, 3994 [4.4]

Grigoryan V. S., Maimistov A. I., Sklyrov Yu. M. (1988), Zh. Eksp. Teor. Fiz. **94**, 174 (English transl.: Sov. Phys. JETP **67**, 530, 1988) [5.2]

Grudinin A. B., Guryanov A. N., Dianov A. M., Zakhidov E. A. (1981), Kvant. Elektron. **8**, 2383 (in Russian) [2.1]

Guzman A., Romagnoli M., Wabnitz S. (1990), Appl. Phys. Lett. **56**, 614 [3.5]

Haake F., Haus J., Schroder G., Glauber R. (1981), Phys. Rev. **23**, 1322 [5.4]

Hammide J. P., Emplit Ph. L., Haelterman M. (1991), Opt. Lett. **16**, 1578 [5.2]

Hasegawa A. (1980), Opt. Lett. **5**, 416 [2.10]

Hasegawa A. (1983), Opt. Lett. **8**, 650 [2.7]

Hasegawa A. (1984), Opt. Lett. **9**, 298 [2.9]

Hasegawa A. (1989), *Optical Solitons in Fibers*, Springer, Heidelberg [1]

Hasegawa A., Brikmann W. F. (1980), IEEE J. Quant. Electr. **16**, 684 [2.8]

Hasegawa A., Tappert F. (1973), Appl. Phys. Lett. **23**, 142 [2.1]

Hasegawa A., Kodama Y. (1981), Proc. IEEE **69**, 1145 [2.6]

Hasegawa A., Kodama Y. (1982), Optics Lett. **7**, 285 [2.1]

Hasegawa A., Kodama Y. (1991a), Phys. Rev. Lett. **66**, 161 [2.7, 3.3]

Hasegawa A., Kodama Y. (1991b), Opt. Lett. **16**, 1385 [2.7, 3.3]

Haus H., Islam M. N. (1985), IEEE Quant. Electr. **21**, 1172 [2.11]

Heatley D. R., Wright E. M., Stegeman G. I. (1988), Appl. Phys. Lett. **53**(3), 172 [3.5]

Hermansson B., Yervick D. (1983), Electr. Lett. **19**, 570 [3.2]

Hida K., Eckern U. (1985), Phys. Rev. B30, 4069 [6.3]

Hodel W., Weber H. P. (1987), Opt. Lett. **12**, 924 [3.3]

Ippen E. P., Shank S. V., Gustavson T. K. (1974), Appl. Phys. Lett. **24**, 190 [2.2]

Ishimaru A. (1978), *Wave Propagation and Scattering in Random Media*, Academic Press, New York [4.5]

Islam M. N., Dijaili S. P., Gordon J. P. (1988), Opt. Lett. **13**, 518 [2.8]

Its A. R., Rybin A. V., Sall M. A. (1988), Teor. Mat. Fiz. **74**, 29 (in Russian) [2.9]

Jain M., Tzoar N. (1978), J. Appl. Phys. **49**, 4649 [2.10]

Jensen S. M. (1982), IEEE J. Quant. Electr. **18**(10), 1580 [3.5]

Kandidov V. P., Shlenov S. A. (1986), Izv. Akad. Nauk SSSR, ser. fiz. **50**, 1191 (in Russian) [4.2]

Karpman V. I., Maslov E. M. (1977), Zh. Eksp. Theor. Fiz. **73**, 537 (in Russian) [2.6]

Karpman I. V., Solov'ev V. V. (1981), Physica D3, 487 [3.1]

Karpman V. I., Maslov E. M., Solov'ev V. V. (1983), Zh. Eksp. Teor. Fiz. **84**, 288 (in Russian) [App. B]

Kath W. L., Ueda T. (1990), Phys. Rev. A42, 563 [2.13]

Kaup D. J., Newell A. C. (1978a), Proc. Roy. Soc. A361, 413 [2.6, 3.6, 4.7]

Kaup D. J., Newell A. C. (1978b), J. Math. Phys. **19**, 798 [2.6, 3.1]

Kaup D. J. (1986), Physica D19, 125 [5.6]

Khikmatov N. A. (1987), Ph. D. thesis, University of Tashkent [2.6, 4.1]

Kitayama K., Sekai S., Ushida N. (1982), Appl. Phys. Lett. **41**, 322 [2.1]

Kivshar Yu. S. (1989), J. Phys. A. **22**, 337 [2.3]

Kivshar Yu. S., Konotop V. V., Sinitsyn Yu. A. (1986), Z. Phys. B **65**, 209 [4.5]

Kivshar Yu. S. (1991), "Perturbation-Induced Dynamics of Dark Solitons in Optical Fibers", in: *Proc. Workshop on Optical Solitons*, ed. Abdullaev F. Kh., World Scientific, Singapore, 46 [2.4]

Klovsky D. D., Sysakyan I. N., Schwartzburg A. B., Shirikov S. M. (1987), Radiotekhnika Elektr. **4**, 740 (in Russian) [4.4]

Klyatzkin V. I. (1980), *Stochastic Equations and Waves in Random Media*, Nauka, Moscow (in Russian) [4.5]

Kodama Y. (1985), J. Stat. Phys. **39**, 597 [5.7]

Kodama Y., Hasegawa A. (1983), Opt. Lett. **8**, 342 [4.6]

Kodama Y., Nozaki K. (1987), Opt. Lett. **12**(12), 1038 [3.3]

Kodama Y., Hasegawa A. (1991),"Generation of Asymptotically Stable Solitons and Suppression of the Gordon-Haus Effect", Preprint A T & T Bell Laboratories, NJ [4.6]

Konotop V. V. (1991), "Dynamics of Multisoliton Optical Pulses with Initial Random Modulations", in: *Proc. Workshop on Optical Solitons*, ed. Abdullaev F. Kh., World Scientific, Singapore, 70 [4.4]

Korepin V. E. (1984), Commun. Math. Phys. **94**, 93 [6.1]

Kosevich A. M., Kivshar Yu. S. (1982), Fiz. Nizk. Temp. **53**, 420 (in Russian) [App. B]

Kravtzov Yu. A., Minchenko A. I. (1984), Kvant Elektron. **11**, 1138 (in Russian) [2]

Krökel D., Halas N. J., Guliani G., Grischkovsky D. (1988), Phys. Rev. Lett. **60**, 29 [2.4]

Krumhansl J. A., Schrieffer J. R. (1975), Phys. Rev. B**11**, 3535 [3.1]

Lam D. K. J., Garside B. K., Hill K. O. (1981), Opt. Lett. **6**, 13 [2.1]

Lamb G. L., Jr. (1980), *Elements of Soliton Theory*, Wiley, New York [5.1]

Lewiz Z. V., Elgin J., Blow K. J., Doran N. J. (1986), "Perturbative Studies of the Zakharov-Shabat Scattering Problem", in: *Dynamical Problems in Soliton Systems*, Springer, Berlin, 30–35 [4.4]

Lichtenberg A. J., Lieberman M. A. (1981), *Regular and Stochastic Motion*, Springer, Heidelberg [4.7]

Lieb E. H., Liniger W. (1963), Phys. Rev. **130**, 1605 [6.0]

Litvak A. G., Talanov V. I. (1967), Radiofizika **10**, 539 (in Russian) [2.8]

Maier A. A. (1982), Sov. J. Quant. Elektron. **12**, 1490 [3.5]

Maimistov A. A., Manykin E. A. (1983), Zh. Eksp. Teor. Fiz. **65**, 1177 (in Russian) [5.4]

Maimistov A. I., Manykin E. A., Sklyarov Yu. M. (1986), Kvant. Elektron. **13**, 2243 (in Russian) [4.4]

Maimistov A. A., Sklyarov Yu. M. (1987), Kvant. Elektron. **14**, 769 (in Russian) [2.4]

Makhankov V. G. (1979), Phys. Rep. **35**, 1 [3.1]

Malomed B. A. (1985), Physica D**125**, 374 [3.3]

Malomed B. A. (1986), Pisma Zh. Tekh. Fiz. **12**, 1419 (in Russian) [2.7]

Malomed B. A. (1986), Teor. Mat. Fiz. **69**, 175 (in Russian) [4.1]

Malomed B. A. (1987), Opt. Commun. **61**, 192 [3.3]

Malomed B. A. (1988), Phys. Scripta **38**, 66 [2.12]

Malomed B. A. (1991), "Inelastic Collisions of Polarized Solitons in a Birefringent Optical Fiber", in: *Proc. Workshop on Optical Solitons*, ed. Abdullaev F. Kh., World Scientific, Singapore, 81 [3.3]

Manakov S. V. (1974), Sov. Phys. JETP **38**, 248 [2.13]

Manykin E. A., Maimistov A. I. (1986), Izv. Akad. Nauk SSSR, ser. fiz. **50**, 1474 (in Russian) [4.4]

Manykin E. A., Surina I. I., Maimistov A. I. (1987), *Propagation of Optical Pulses in Fibers* (Review), Inst. of Atomic Energy, Moscow [1]

Marcuse D. (1972), *Light Transmission Optics*, Van Nostrand Reinhold, New York [App. A]

Matinyan S. G., Savvidi G. K. (1982), JETP Lett. **34**, 590 [2.13]

McCall S. L., Hahn E. L. (1967), Phys. Rev. Lett. **18**, 908 [1, 5.1]

McLaughlin D. W. (1981), J. Stat. Phys. **24**, 377 [4.7]

McLaughlin D. W., Scott A. C. (1978), Phys. Rev. A**18**, 1652 [App. C]

Menyuk C. R. (1986), Phys. Rev. A**33**, 4367 [2.13]

Menyuk C. R. (1987), Opt. Lett. **12**, 614 [2.12]

Meyer A. A., (1982), Kvant Elektron. **9**, 2296 (in Russian) [3.5]

Meyer A. A. (1985), Kvant Elektron. **13**, 1360 (in Russian) [3.5]

Mitschke F. M., Mollenauer L. F. (1986), Opt. Lett. **11**, 659 [2.12]

Mollenauer L. F. (1988), in: *Structure, Coherence and Chaos in Dynamical Systems*, Manchester University Press, 257–286 [1]

Mollenauer L. F., Stolen R. H., Gordon J. P. (1980), Phys. Rev. Lett. **45**, 1095 [2.1, 4]

Mollenauer L. F., Stolen R. H. (1982), Fiber Optics Technology **4**, 193 [2.4]

Mollenauer L. F., Stolen R. H. (1982), Laser Focus **18**, 193 [2.4]

Mollenauer L. F., Stolen R. H. (1984), Opt. Lett. **9**, 13 [2.11]

Mollenauer L. F., Smith K. (1988), Opt. Lett. **13**, 675 [2.11]

Mollenauer L. F., Smith K. (1989), Opt. Lett. **14**, 1284 [3.2]

Moloney J. V. (1985), IEEE J. Quant. Electr. **21**, 1393 [4.8]

Moloney J. V. (1986), Phys. Rev. A33, 4061 [4.8]

Mork J., Christiansen P. L., Lassen H. E., Tromborg B. (1987), IEEE Proc. **134**, pt.J, 127 [4.4]

Muraki D. (1990), Ph. D. thesis, Northwestern University, Chicago [2.13]

Nesterova Z. V., Aleksandrov I. V. (1985), Zh. Eksp. Teor. Fiz. **88**, 96 (in Russian) [2.5]

Newell A. C. (1985), *Solitons in Mathematics and Physics*, SIAM, Arizona [2.7]

Nozaki K., Bekki N. (1983), Phys. Rev. Lett. **50**, 1226 [4.7]

Okhura K., Ichikawa Y. H., Abe Y. (1987), Opt. Lett. **12**, 516 [2.6]

Ostrovsky L. A. (1959), Radiofizika **2**, 833 (in Russian) [5.3]

Ostrovsky L. A. (1963), Zh. Tekh. Fiz. **33**, 905 (in Russian) [2.3]

Ostrovsky L. A. (1966), Zh. Eksp. Teor. Fiz. **51**, 1189 (in Russian) [1, 2.5]

Potasek M. J. (1987), Opt. Lett. **12**, 717 [2.8]

Potasek M. J., Agrawal G. P. (1986), Electron. Lett. **22**, 759 [2.8]

Prokhorov A. M. (1983), Izv. AN SSSR **47**, 1874 (in Russian) [2]

Pikovsky A. S. (1985), "Stochastic Breaking of Solitons due to Chaotic Motion", in *Waves and Diffraction*, Tbilisi State University Press, 154–156 (in Russian) [4.7]

Rajarman P. (1982), *Solitons and Instantons in Quantum Field Theory*, North-Holland, Amsterdam [1, 6.1]

Reynaud F., Barthelemy A. (1990), Europhysics Lett. **12**, 401 [3.2]

Raizer Yu. P. (1974), *Laser Spark and Propagation of Spark*, Nauka, Moscow (in Russian) [5.7]

Romagnoli M., Wabnitz A., Zoccolotti L. (1991), Opt. Lett. **16**, 1249 [3.5]

Rothenberg A. B., Heinrich H. K. (1992), Opt. Lett. **17** 261 [2.3, 4]

Rupasov V. I. (1982), Kvant Elektron. **9**, 2127 (in Russian) [5.3]

Rupasov V. I. (1983), Zh. Eksp. Teor. Fiz. **83**, 1711 (in Russian) [6.4]

Rupasov V. I., Yudson V. I. (1984), Zh. Eksp. Teor. Fiz. **84**, 818 (in Russian) [6.3]

Rytov S. M. (1976), *Introduction to Statistical Radiophysics*, Nauka, Moscow (in Russian) [4.5]

Satsuma J., Yajima K. (1974), Suppl. Prog. Theor. Phys. **55**, 284 [2.3,4.4]

Shukla P. K., Rasmussen J. J. (1986), Opt. Lett. **11**, 171 [2.8]

Sinai Y. G. (1966), Izv. Akad. Nauk SSSR, Math. **30**, 15 (in Russian) [4.7]

Sklyanin E. K. (1978), *The Inverse Scattering Theory and the Quantum NSE*, Preprint Leningrad Dept. of Math. Inst., Leningrad [6.2]

Sklyanin E. K. (1979), *On Complete Integrability of the Landau-Lifshitz Equation*, Preprint Leningrad Dept. of Math. Inst., Leningrad, E-3-79 [6.2]

Snyder A. U., Love J. D. (1981), *Optical Waveguide Theory*, Chapman and Hall, London [App. A]

Steudel H. (1989), *Inverse Scattering Theory of Superfluoresence*, Preprint Inst. Opt. Spectr. 89-2, Berlin [5.4]

Stolen R. H. (1979), IEEE J. Quant. Electron. **15**, 1157 [2]

Stolen R. H. (1981), Opt. Lett. **6**, 213 [2]

Stolen R. H., Leibolt V. N. (1976), Appl. Optics **15**, 239 [2.1]

Stolen R. H., Lin C. (1978), Phys. Rev. A17, 1448 [2.2]

Sysakyan I. N., Schwartzburg A. B. (1984), Kvant Elektron. **11**, 1703 (English transl.: Sov. J. Quant. Electron. **14**, 1146, 1984) [1]

Tadjimuratov S., Tartakovsky G. (1987), Dokl. Akad. Nauk UzSSR **5**, 32 (in Russian) [5.2]

Tai K., Hasegawa A., Tomita A. (1986), Phys. Rev. Lett. **56**, 135 [2.8]

Tajima K. (1987), Opt. Lett. **12**, 54 [1, 2.7]

Talanov V. I. (1964), Radiofizika **7**, 564 (in Russian) [1]

Thacker H. B. (1981), Rev. Mod Phys. **53**, 258 [6.2]

Tomlinson W. J., Hawkins R. J., Weiner A. M., Heritage J. P., Thurston R. N. (1989), J. Opt. Soc. Am. B**6**, 329 [2.4]

Tracy E.R., Chen H., Lee Y. C. (1984), Phys. Rev. Lett. **53**, 218 [2.9]

Tracy E. R., Chen H. H. (1988), Phys. Rev. A**37**, 815 [2.9]

Tracy E. R., Lasson I. W., Osborne A. R., Bergamasco L. (1988), Physica D **32**, No. 1, 83 [2.5]

Tratnik M. V., Sipe J. E. (1988), Phys. Rev. A**38**, 2011 [2.13]

Tratnik M.V. (1992), Opt. Lett. **17**, 917 [2.13]

Trillo S. et al. (1989), Opt. Commun. **70**, 166 [2.13]

Trillo S., Wabnitz S., Wright E. M., Stegeman G. I. (1988), Opt. Lett. **13**, 871 [3.5]

Tzoar N., Gersten J. I. (1976), *Optical Properties of Highly Transparent Solids*, eds. Miura S., Bendow B., Plenum, New York [2.1]

Umarov B. A. (1986), Ph. D. thesis, Thermophysics Dept. of the Uzbek Acad. Sci., Tashkent (in Russian) [4.5]

Unger H.-G. (1977), *Planar Optical Waveguides and Fibers*, Clarendon, Oxford [App. A]

Vysloukh V. A. (1982), Usp. Fiz. Nauk **136**, 519 (in Russian) [1, 2.1, 3.2]

Vysloukh V. A., Serkin V. N. (1983), Pis'ma Zh. Eksp. Teor. Fiz. **38**, 170 (English transl.: JETP Ltt. **38**, 199, 1983) [2.6]

Vysloukh V. A., Serkin V. N. (1984), Izv. Akad. Nauk SSSR **48**, 1777 (in Russian) [2.1]

Vysloukh V. A., Matveeva G. A. (1985), Radiofizika **28**, 101 (in Russian) [2.1, 4.2]

Vysloukh V. A., Cherednik I. V. (1986), Dokl. Akad. Nauk SSSR **289**, 336 (in Russian) [3.2]

Vysloukh V. A., Ivanov A. V. (1988), Izv. Akad. Nauk SSSR, ser. fiz. **52**, 359 (in Russian) [4.4]

Vysloukh V. A., Sukhotskova N. A. (1988), Pisma Zh. Eksp. Teor. Fiz. **14**, 818 (in Russian) [4.4]

Wabnitz S. (1988), Phys. Rev. A**38**, 2018 [2.13]

Wabnitz S., Trillo S., Wright E. M., Stegeman G. I. (1991), JOSA B**8**, 602 [3.5]

Wai P. K., Menyuk C., Chen H. H., Lee Y. C. (1987), Opt. Lett. **12**, 628 [2.6]

Wai P.K., Menyuk C.R., Chen H.H. (1991), Opt. Lett. **16**, 1231 [4.5]

Wright E. M., Stegeman G. I., Wabnitz S. (1989), Phys. Rev. A**40**, 4455 [3.5]

Yervick D., Hermanson B. (1983), Opt. Commun **47**, 101 [2.1]

Zakharov V. E. (1971), Zh. Eksp. Teor. Fiz. **60**, 993 (in Russian) [3.1]

Zakharov V. E., Berkhoer A. L. (1970), Sov. Phys. JETP **31**, 486 [2.13]

Zakharov V. E., Shabat A. B. (1971), Zh. Eksp. Teor. Fiz. **61**, 26 (in Russian) [2.1, 2]

Zakharov V. E., Manakov S. V., Novikov S. P., Pitaevsky L. P. (1984), *Theory of Solitons*, Consultants Bureau, New York [2.3, 3.2, 4.3, 5.1]

Zakharov V. E., Gabitov E., Mikhailov A. V. (1984), Zh. Eksp. Teor. Fiz. **86**, 1204 (in Russian) [5.3]

Zamolodchikov A. B. (1979), Commun. Math. Phys. **69**, 165 [6.2]

Zaslavsky G. M. (1984), *Stochasticity of Dynamical Systems*, Nauka, Moscow (in Russian) [4.7]

Ziman J. (1979), *Models of Disorder*, Cambridge University Press [5.1]

Zysset B., Beaud P., Hodel W., Weber H. P. (1986), in *Ultrafast Phenomena V*, Springer, Berlin, p.54 [2.11]

Subject Index

Adiabatic approximation 27, 77, 97
Amplification 30
– random 101, 103
– of solitons 30
Amplifiers 102
– gain of 102, 105
– noise 105
Anomalous dispersion 2, 20, 31
Antisoliton 96, 169
Autocorrelator 21
Autosolitons 4, 115, 199, 121

Beam 1, 111
Bethe ansatz 139, 148
Binding energy 62
Birefringence 50, 54, 101
Breather 106, 169
– quantum 151
Brillouin scattering 5, 13

Carrier wave 58
Chirikov criterion 109
Cladding 161
Coefficient
– amplification 102
– coupling between fibers 76
– Jost 35, 89
– linear absorption 110
– scattering 136
Color center laser 21, 50
Communication system 3, 6
Commutational relationship 144, 147, 154
Component
– Jost 69
– Stokes 31
Compression, nonlinear 13
Conservation laws 27, 32
Correlation
– function 81, 87, 90, 98
– length 96, 100
Coupling
– constant 150
– nonlinear, between modes 46

Critical peak power 10
Cut-off frequency 162

Dark soliton 20, 23, 166
Damping
– coefficient 133
– of inversion 116
– of pulse 10
Density matrix 125
Dicke model 152
Dipole
– approximation 152
– moment 115
Dispersion
– group delay 34
– higher-order 27, 37, 66
– intermode 5, 48
– length 121
– linear 7
– material 1, 5
– normal 20
– second-order 23
– waveguide 5
Dissipative perturbation 70
Distortion parameter 66
Dynamical chaos
– optical solitons 105, 106
– soliton 4
Dynamical stochasticity 81

Electromagnetic
– beam 108
– fields 2, 106, 152
– shock waves 23
Electrostrictional mechanism 65
Envelope soliton 106
Equation
– Bessel 161
– Bloch 115
– Fokker-Planck 102
– Korteweg-de Vries 1, 96
– Marchenko 128, 136
– Maxwell-Bloch 115, 125, 154

– Nonlinear Schrödinger (NLS)   10, 13, 29, 164
– of motion   61
– sine-Gordon (SG)   99, 101, 126, 169

Fiber
– active   101, 119
– birefringent   71
– capillary   11, 26
– core   12
– cross-section   10
– doped with ND   121
– gradient   5
– multi-mode   46
– nonuniform   97
– single-mode   6, 12, 162
– step-index   47
– tunnel-coupled   76
– with resonant impurities   119
Finite gap
– potential   173
– solution   71, 125
Frequency
– atomic transition   115
– detuning distribution function   116
Fresnel number   110
Function
– Gain   120
– Green   86
– Jost   89, 132, 134, 145
– Lagrange   58, 86
– random Gaussian   128

Group dispersion
– negative   13
– positive   11, 95
– velocity   47, 130

Hamiltonian
– of breather   170
– NLS   140, 144, 165
– SG   170

Induction
– electric   7
– nonlinear part   7
Integral invariants
– NLS   95, 165
Interaction
– phase independent   65
Interferometer
– Fabry-Perot   12, 110
– Michelson   64
Inverse scattering transform (IST)   4, 18, 27,
    118, 131

Kerr nonlinearity   2, 6, 119
Kink   60, 96

Level degeneracy   127

Mass, effective   60
Media
– active   115
– condensed   118
– nonstationary   108
– resonant   115
– three-level   122
Method
– characteristics   24
– direct   58, 63
Medium
– nonlinear dispersive   15
– resonant   2, 115
Modulational instability   3, 37
Molecular relaxation   133
Monodromy matrix   141
Motion integral   143, 151
Multisoliton state   159

N-soliton solution   19, 164
N-string   149
Nearly integrable system   62, 66, 81
Noise signal   86
Nonlinear Fraunhofer diffraction   88
Nonlinear periodic wave   71
Nonlinear resonance overlapping   109

Optical bistability   110

Parametric resonance   109
– stochastic   104
Permittivity   7
Perturbation theory   27, 61, 166
Phase
– noise   92
– random modulation   81, 96
– screen   97
– self modulation   11
Poisson brackets   140
Population inversion   121
Potential
– effective   60
– initial random   91, 96
– interaction   60, 78
Propagation constant   46
Pump wave   30
Pulse
– broadening   5, 11, 22
– length   12, 123

- narrowing   13
- random   81, 88
- self-action   87
- short   22
- spectrum   22
- 2π 119, 123

Quantum
- IST   139
- field model   139
- nonlinear Schrödinger equation   139, 144
- sine-Gordon system   149

Raman amplification   3, 34
- gain   34
- -spectrum   53
- self-frequency shift   65
Refraction index   9, 76
- nonlinear   10
Resonant impurities   115, 119
Resonator
- optical ring   110
Riemann
- surface   75
- θ-function   72, 173
Rotating wave approximation   153

Saturation   120
Self-focussing   1, 11
- resonance pulse   118
- threshold   8
Self-induced transparency   118, 132
Self-similarity   129
Self-steepening   25, 38, 66
Separatrix   43, 105
Solitary wave   14, 58
Soliton
- beam   107
- bright   2
- chaotic motion   109
- coupler   3
- envelope   1
- fundamental dark   20
- generation   84, 90, 91
- gray   20
- higher-order   23
- interactions   51, 63

- -elastic   3
- -inelastic   3
- laser   48, 64, 159
- multicomponent   3, 54
- number   20
- period   22
- quantum   139
- self-frequency shift   52
- spatial   1, 65
- SRS   130, 133
- switching   76
- vector   54
Spectral parameter   89, 95, 120
- pattern   71
Stimulated Raman scattering   104, 130
- in fiber   30
Stochastic layers   105
Superfluorescence   118, 125
Superradiance   152

Three-level atoms   122
Topological charge   169
Topological soliton   1
Turbulence, optical   110
Two-coupled fibers   3, 56, 76
Two-levels atoms   115, 119, 153
Two-soliton state   62, 66

Variables
- action-angle   84

Wave
- cnoidal   15, 71
- envelope   1, 14
- linearly-polarized   37
- packet   109
- plane   37
- self-modulation   39
- Stokes   131, 133
Waveguide
- optical   3
- two tunnel-coupled   76

Yang-Baxter relation   149

Zakharov-Shabat system   89, 118, 144